Hot-Cat 2.0

How last generation E-Cats are made

Eng. Roberto Ventola

In collaboration with Vessela Nikolova

This book is dedicated to the memory of Sergio Focardi (1932-2013),
Physicist and Professor Emeritus at the University of Bologna.

CONTENTS

Preface   1

Introduction   3

Chapter 1 - The Hot-Cat "Type I design"   7

Chapter 2 - The Hot-Cat "Type II design"   11

Chapter 3 - The Hot-Cat "Type III design"   17

Chapter 4 - The Hot-Cat "Type IV design"   25

Chapter 5 - A possible alternative design   33

Chapter 6 - The secret interior of a reactor   47

Chapter 7 - Why the Hot-Cat works   55

Chapter 8 - How the E-Cat HT was invented   65

Chapter 9 - Salient features of a Hot-Cat   73

Chapter 10 - Unveiling the "Cat & Mouse" design   83

Chapter 11 - Two types of long-term stability   95

Chapter 12 - Control of Hot-Cat reactions   105

Chapter 13 - Two underestimated problems   119

Chapter 14 - A solution for the powder charge   129

About the Authors   139

PREFACE

I want to be absolutely sincere: this book was written by Eng. Roberto Ventola, who has a technical background, and I appear undeservedly as an author. Therefore, I want to tell the reader how things really went.

Seeing as Eng. Ventola sent many technical articles to be published on the blog and I could not publish them all together, I suggested collecting and publishing them in a book, an idea that he enthusiastically agreed on, offering – in exchange for this opportunity – to donate half of the earnings to Vessy's Blog Fund through my co-authorship.

Since my blog is financed only with contributions coming from generous readers, and because this is essential for its future survival, I have agreed to appear as co-author. However, my real contribution was only to standardize the style of the book to that of previously published articles and to make a few public communications.

I am very happy to have contributed in some way to the birth of this work, which I think will remain for many years as a reference book for all those who are seriously interested in the last generation Hot-Cat. It is a bit more technical than my book "E-Cat - The New Fire", but I'm sure that most of readers will be able to appreciate it.

Despite the fact that I am not able to assess its more technical content, it nevertheless seems very intellectually stimulating reading to me, enhanced by about 100 figures that enrich the text, making it almost unique.

I hope that this is really the beginning of a world energy revolution, though at this point many new characters will need to appear on the scene.

Vessela Nikolova

INTRODUCTION

I think there is no better introduction to this book than the words written by Andrea Rossi in a recent comment posted on the blog of his *Journal of Nuclear Physics* (JoNP) in response to a reader who asked if "it could be possible to make an industrial or domestic E-Cat, working every day for domestic or industrial customers, without using catalyzers, but only the fuel described in the Lugano Report":

JC Renoir:

No, it is impossible. To make an E-Cat work regularly for months you need more than that. As I said, the E-Cat is a **much more complicated thing than is commonly imagined**. The substantial underestimation of what we have done has given us a real advantage, since instead of changing the game, the attempts that are going on to imitate us only try and fix **old schemes**, thinking that, if we have done anything that makes a difference, it must be something very small, close to evanescent. This attitude has given us a great advantage in the competition. For years I have been considered an imbecile who has just made it lucky, God knows why, and who creates things without understanding what the heck he is doing (in the best of cases). Either that or, more kindly, I am considered a fraudster. Obviously this has been a **big advantage** for our Team. I can say this now, since we are close to going commercial on a massive scale. I think it is now too late to catch up on us.

Warm Regards,
A.R.

Yes, reading this book the reader will understand perfectly how the new "Hot-Cat 2.0" – as I call the E-Cat of the latest generation – is a much more sophisticated apparatus than you might imagine from an its superficial analysis. And if you do not understand these "superior" characteristics, even a "Parkhomov-like" replication is totally useless. In fact, it would only be a starting point and by no means the point of arrival.

Therefore, this book is not aimed at those who want to replicate the old generation Hot Cat, although it would be extremely useful for that purpose as well. In any case, it is aimed primarily at those who, having already understood that kind of E-Cat, want to move on to an industrial product, or, that is to say, a second-generation E-Cat. Rossi has spent quite a few years carrying out this crucial development, and in this book you will find the evolution of his invention.

Rossi is also right in saying that many people still think of the E-Cat with old patterns, because the evolution of this product has led from what was apparently a rudimentary prototype to a small gem of science and technology. In addition, there are huge differences between an original "low temperature" E-Cat and a modern Hot-Cat, which works at high temperatures. This book will help you to understand all of these in detail.

"Hot-Cat 2.0" will show you a sort of "naked" Hot-Cat, virtually passing the device through an X-ray and giving a look at it from completely new points of view. You'll find that almost all the mysteries and inconsistencies noted by many observers during the Third-Party testing are perfectly solved if you accept the fact that an E-Cat is now a complex machine whose intellectual property must be protected in every way.

By making a careful analysis of all – and I really mean all – the publicly available information (which has been carefully preserved and interpreted) we will be – at least partially – ripping off the veil of secrecy surrounding an E-Cat, and telling the story of this invention-cum-product, thus allowing you to enter a sort of "time machine". In this way, you'll understand why Rossi and his team have chosen certain roads and not others in their attempts to obtain the best possible results.

As Rossi points out, the advantage gained by his team is now remarkable, thanks to a substantial underestimation of their work by the audience who is passionate about this story. On the other hand, it is also true that the information made public last autumn allows for a partial replication of the Rossi Effect and of a Hot-Cat, though all this is still insufficient by itself to create a commercially viable product.

The first step to be taken in order to replicate a Hot-Cat 2.0 is to understand how it is done. This is not always easy, for the simple reason that Rossi had sometimes had fun misdirecting people. In this long analysis, you will discover when this happened, and how, at the same time, he has also on occasions provided crucial information in a number of oral interviews. Perhaps, such important intellectual property could (and should) be better protected.

I would like to thank Vessela for giving me this unique and unexpected opportunity. I hope you will like this book and that it gives you a real leap forward in your understanding of the Hot-Cat. If you are interested in contacting me – for example, to report any inaccuracies – please email me to: roberto.ventola@yahoo.com.

Thank you in advance!

Eng. Roberto Ventola

CHAPTER 1

THE HOT-CAT "TYPE I DESIGN"

The picture here below, published in a skinny version in the patent application filed by *Industrial Heat* on April 26, 2014, shows a layered tubular **reactor device** (Fig. 1 in the cited document), also represented in cross-sectional view (Fig. 2). It can be described as *Energy Catalyzer HT*, where HT stands for "high temperature" and it is **the first of three** different **embodiments** described in such patent application, so for sake of simplicity hereinafter I'll indicate it as "*E-Cat HT – I*".

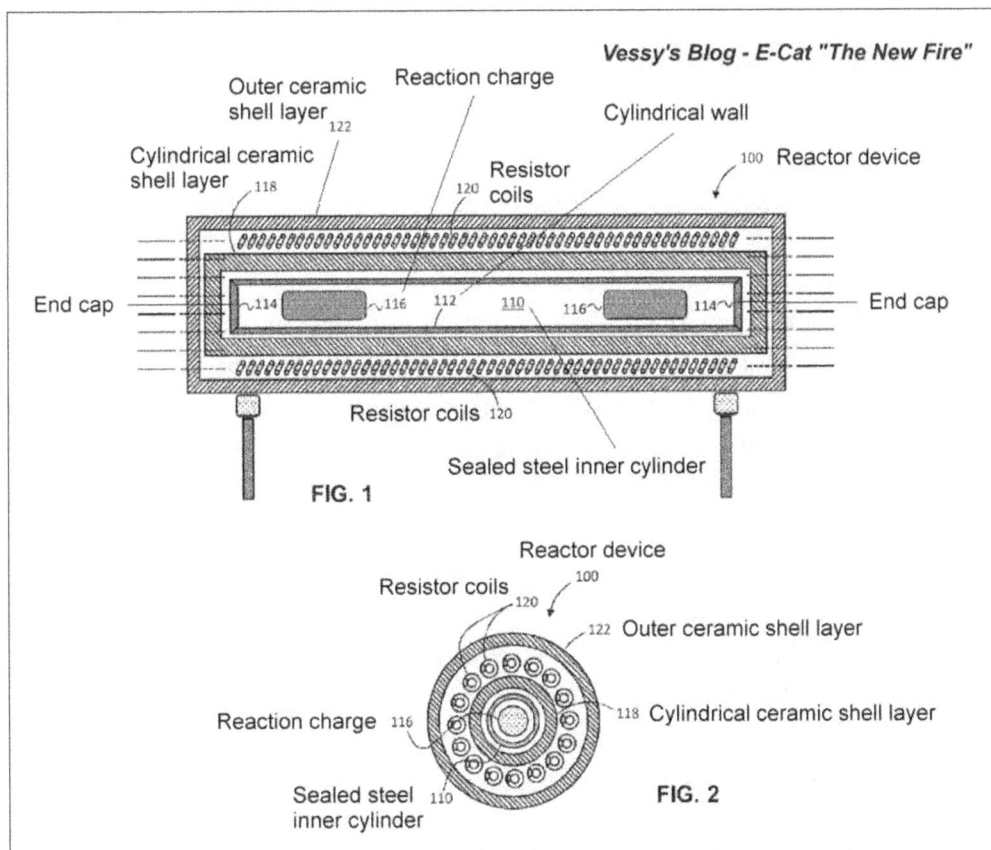

Fig. 1.1 - Diagram of a reactor device E-Cat HT "Type I design" (from IH's patent, slightly modified).

This reactor, with the charge non evenly distributed but concentrated in two distinct locations along the central axis of the reactor, **was used in the first of the three tests** described in the first *Third Party Report* (TPR-1), performed in November 2012 (see the table about the TPR tests on Hot-Cat reactors). Such test failed, due to the **overheating** and melting of the steel cylinder containing the active charge and the surrounding ceramic layers.

Document	TPR-1	TPR-1	TPR-1	TPR-2
Test date	Novemb. 2012	Decemb. 2012	March 2013	March 2014
Location	Ferrara (IT)	Ferrara (IT)	Ferrara (IT)	Lugano (CH)
Reactor type	E-Cat HT - I	E-Cat HT - II	E-Cat HT - III	E-Cat HT - IV
Temperature	793 °C	436 °C	302 °C	1260-1400 °C
Fuel	Ni, H, ?	Ni, H, ?	Ni, H, ?	Ni, H, Li, ?
Duration	Failed	96 hours	116 hours	32 days
Self-Sustained	No	No	Yes	No
COP	not available	5.6 +/- 0.8	2.9 +/- 0.3	3.2-3.6

Table 1.1 - All the tests described in the Third Party Reports released from the scientists Levi et al.

According to the description given in the cited patent and integrating the info contained on this issue in TPR-1, a **sealed steel inner tube** (110) included a cylindrical wall (112) that extended between two end caps (114). The inner tube contained **reaction charges** (116) **in two** distinct longitudinal **locations**. A first cylindrical ceramic shell layer (118) surrounded the inner tube.

Each of **16 resistor coils** (120) extended the length of the interior of the reactor device between the inner cylindrical ceramic shell layer and a more **outer cylindrical ceramic shell** layer. The resistor coils were circumferentially distributed around the inner cylindrical ceramic shell layer to produce **uniformly distributed heating** when electrical current was passed through the coils.

According to the patent application, the resistor coils were operated continuously at **about 1 kW** to perform experimental investigations of heat production. Once operating temperature is reached, it is possible to **control the reaction by regulating the power** to the coils.

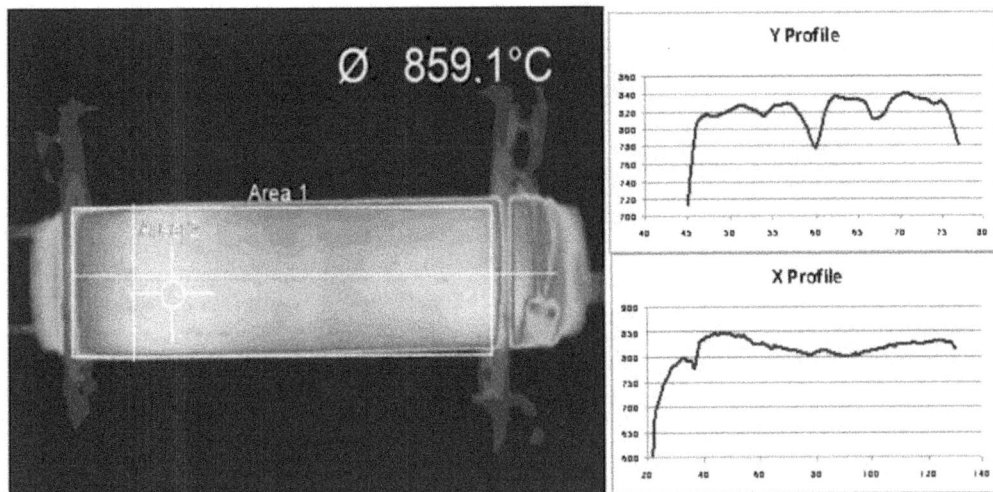

Fig. 1.2 - An IR thermal image of the November 2012 test device. Area 1 is at 793 °C. The temperature dips visible in the diagram on the right are shadows of the resistor coils, projected on the IR thermal camera lens by a source of energy of higher intensity located inside the device.

The reactor device was charged with a small amount of **hydrogen loaded nickel powder**. However the fuel was, more precisely, a mixture of nickel, hydrogen **and a catalyst** consisting, according to TPR-1, of some **"additives"** pressurized with the hydrogen gas and not disclosed being an industrial trade secret.

The E-Cat HT-I is a further **high-temperature development** of the original apparatus described in detail in the old patent application *WO 2009125444*, which has also undergone many changes in the last years. As in the original E-Cat, the powder charge activated by heat produced by the resistor coils produces **excess heat** from some type of reaction.

As said before, unfortunately the reactor was **destroyed** in the course of the experimental run.

Before melting, it looked just like in the picture below, where you can see the shining charges distributed laterally in the reactor and the **horizontal darker lines**, corresponding to the shadows of the resistor coils, projected outward by a source of thermal energy located further inside the device, and **of higher intensity** as compared to the energy emitted by the coils themselves.

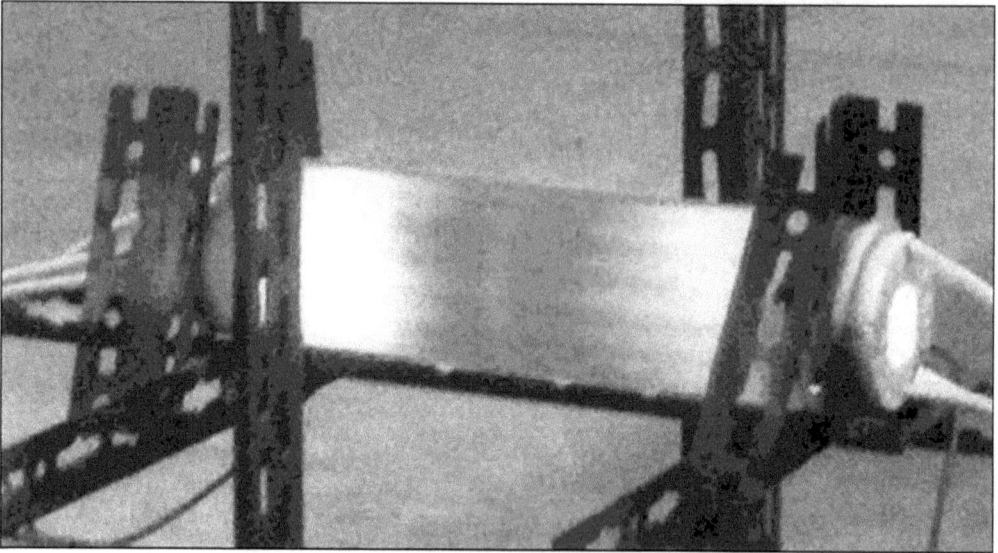

Fig. 1.3 - The E-Cat HT "Type I design" during the Third Party test performed on November 20, 2012.

Therefore, this is **evidence of an exothermic reaction** that occurred within the inner tube.

The test was fruitful also because it demonstrated in a more direct way, i.e. completely destroying the entire reactor, a **huge production of excess heat**, which however could not be quantified. The device had similar, but not identical, features to those of the reactors used in the December 2012 and March 2013 TPR-1's runs, which I'll illustrate in detail in my future contributes.

CHAPTER 2

THE HOT-CAT "TYPE II DESIGN"

The picture here below, published in a skinny version in the patent application filed by *Industrial Heat* on April 26, 2014, shows a layered tubular **reactor device** (Fig. 4 in the cited document), also represented in cross-sectional view (Figg. 5 and 6). It is **the second of three** different **embodiments** described in such patent application, so hereinafter I'll indicate it as "*E-Cat HT – II*".

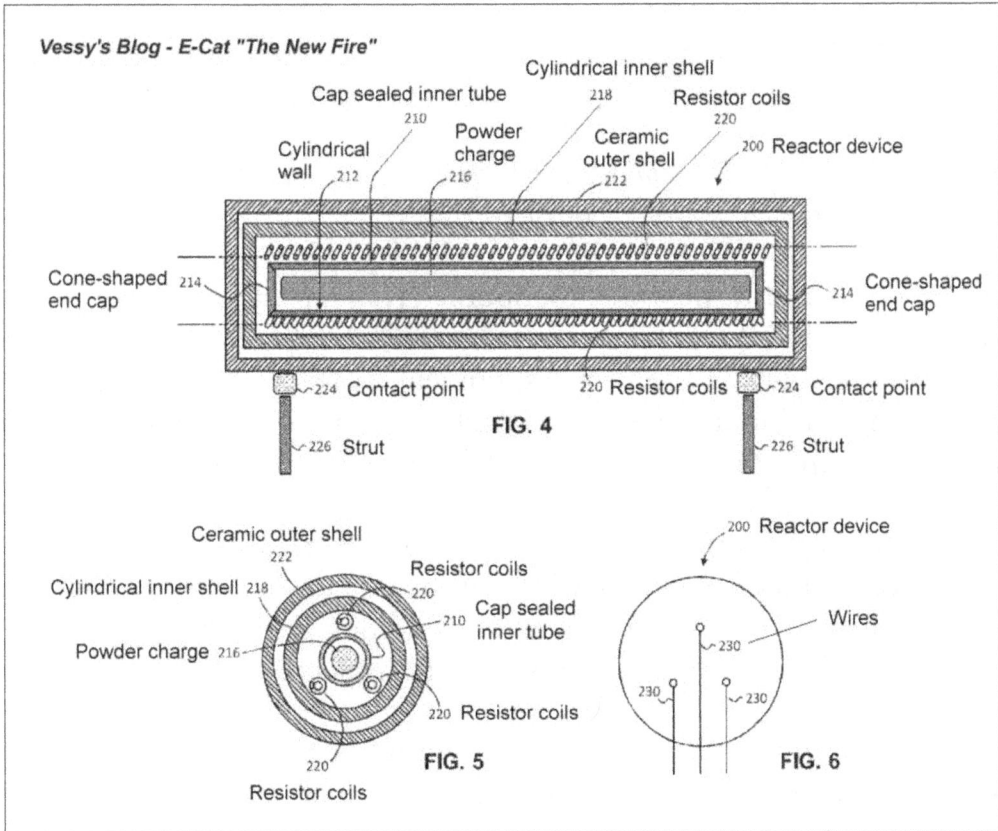

Fig. 2.1 - Diagram of a reactor device E-Cat HT "Type II design" (from IH's patent, slightly modified).

This reactor, with the powder charge widely and uniformly distributed along the central axis of the reactor, **was used in the second of the three tests**, or "experiments", described in the first *Third Party Report* (TPR-1). Such experiment consisted in a 96-hours run of the device **continuously powered** – i.e. never operating in self-sustained mode – and was performed, successfully, on December 13-17, 2012 in Ferrara, Italy.

Document	TPR-1	TPR-1	TPR-1	TPR-2
Test date	Novemb. 2012	Decemb. 2012	March 2013	March 2014
Location	Ferrara (IT)	Ferrara (IT)	Ferrara (IT)	Lugano (CH)
Reactor type	E-Cat HT - I	E-Cat HT - II	E-Cat HT - III	E-Cat HT - IV
Temperature	793 °C	436 °C	302 °C	1260-1400 °C
Fuel	Ni, H, ?	Ni, H, ?	Ni, H, ?	Ni, H, Li, ?
Duration	Failed	96 hours	116 hours	32 days
Self-Sustained	No	No	Yes	No
COP	not available	5.6 +/- 0.8	2.9 +/- 0.3	3.2-3.6

Tab. 2.1 - All the tests described in the Third Party Reports released from the scientists Levi et al.

According to the dispersive description given in the cited patent and widely integrating the information contained on this issue in TPR-1, the reactor device (200) used in this experiment was a layered cylindrical device having an **inner tube** (210). Such inner tube, made of AISI-310 steel, had a 3 mm thick cylindrical wall (212) with a 33 mm diameter.

Two cone-shaped **end caps** (214) made of **AISI-316** steel were hot-hammered into the longitudinal ends of the inner tube, sealing it hermetically. Cap adherence was obtained by exploiting the higher thermal **expansion coefficient** of AISI-316 steel with respect to AISI-310.

As such, the inner tube constitutes a **vessel sealed** against ingress or egress of matter, including gaseous hydrogen. This represents a distinction of this type of reactor over previous reaction vessels (normal E-Cat or, if you prefer, E-Cat LT, where LT stands for "Low temperature"), that were preloaded with **pressured gases** such as hydrogen (see the previous Patent Application *WO 2009125444*, international extension of an Italian patent filed in 2008).

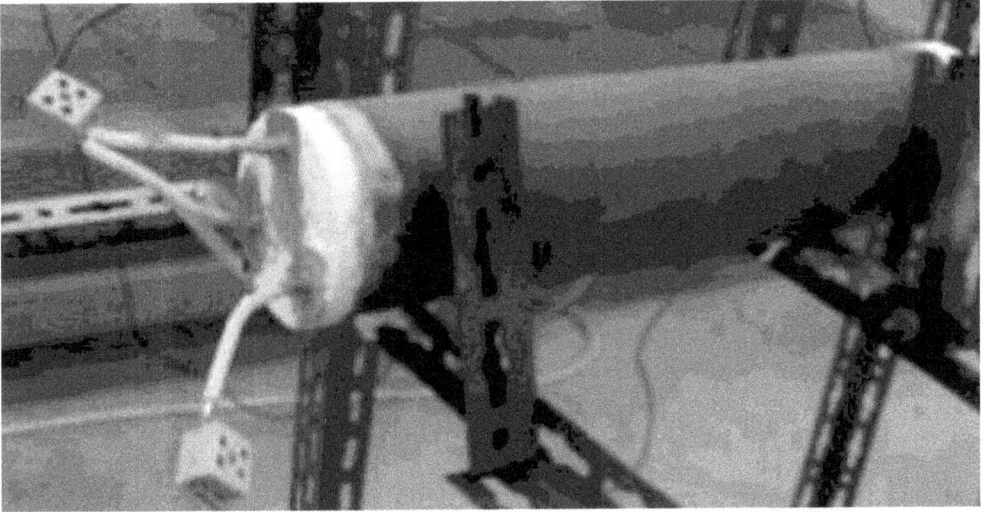

Fig. 2.2 - The E-Cat HT "Type II design" before the Third Party test performed on December 13-17, 2012. You can see the black paint and the power cables to the three internal resistor coils.

The inner tube contained a powder **reaction charge** (216) uniformly distributed along the axis of the device, and consisting of a small amount of **hydrogen loaded nickel powder**. However the fuel was, more precisely, a mixture of nickel, hydrogen **and a catalyst** consisting, according to the TPR-1, of some "**additives**" pressurized with the hydrogen gas and not disclosed being an industrial trade secret.

A **silicon nitride** cylindrical **outer shell** (222), 33 cm in length and 10 cm in diameter, was coated with a special aeronautical-industry grade **black paint** (produced in the N-E of Italy), capable of withstanding temperatures up to 1200 degrees Celsius. A cylindrical inner shell (218), which was made of different ceramic material – **corundum** – was located within the outer shell.

The inner shell housed **three delta-connected** spiral-wire **resistor coils** (220), which were laid out horizontally, parallel to and equidistant from the center axis of the device. The three resistor coils essentially run the interior length of the device and were **independently** wired to a power supply by wires (230) that extended outward from the reactor device (see Fig. 6).

The resistor coils within the reactor were fed by a **Triac power regulator** device (302, see Fig. 7) which interrupted each phase periodically, in order to modulate the power input with a controlled waveform, which is an industrial trade **secret waveform**. This procedure, needed to **properly activate** the powder reaction charge, had no bearing on the power consumption of the device, which remained constant throughout the experiment.

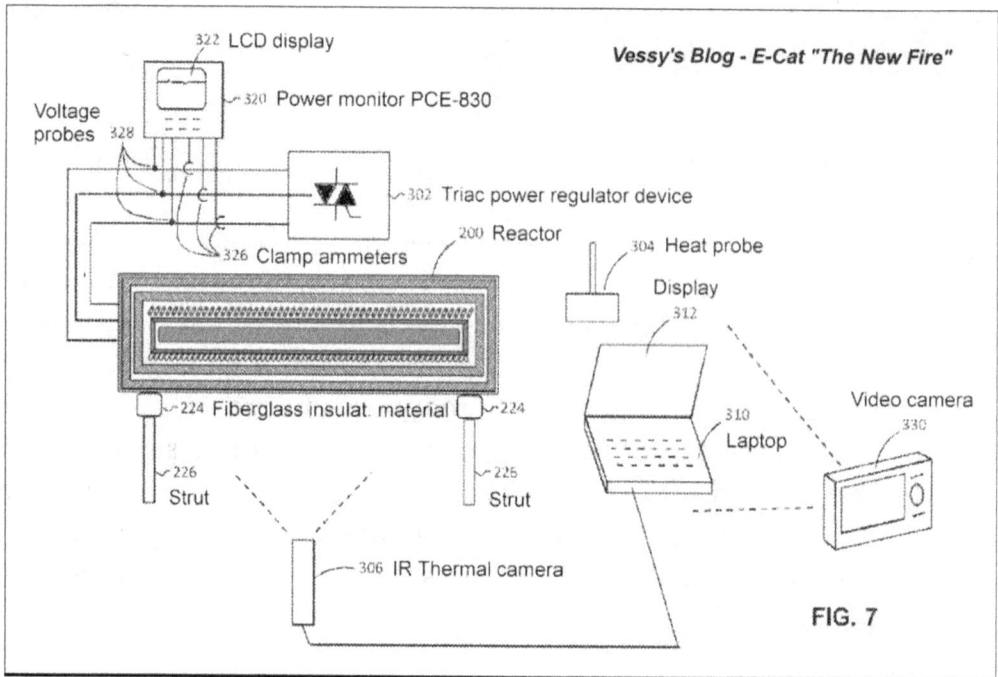

Fig. 2.3 - The experimental setup of the second test on a Hot-Cat reactor described in this article.

Due to the failure in the first test performed in November 2012, when the primer resistor coils were run at about 1 kW, in this second experiment the continuous **power input** to the reactor was limited to a much lower value, **360 W**, so the E-Cat HT's hourly power consumption was 360 W. The E-Cat HT's power production was **almost constant**, with an average of 1609 W (Fig. 8).

A wide band-pass **power quality monitor** (320) – a *PCE-830 Power and Harmonics Analyzer* produced by PCE Instruments – measuring the electrical quantities on each of the three phases was used to record the power absorbed by the resistor coils. It was connected directly to the reactor device resistor coil power cables by three **clamp ammeters** (326) and three **probes** (328), respectively for current and voltage measurements.

Finally, an **IR thermography camera** (306), model *Optris PI Thermal Imager*, was used to acquire a thermal image on a display (312) and to measure the surface temperature of the reactor device with a **2% precision** of measured value, in order to make an **infrared thermographic calorimetry**. The thermal camera was positioned about 70 cm below the reactor device in order not to damage the camera itself from the heat transferred by rising convective air currents.

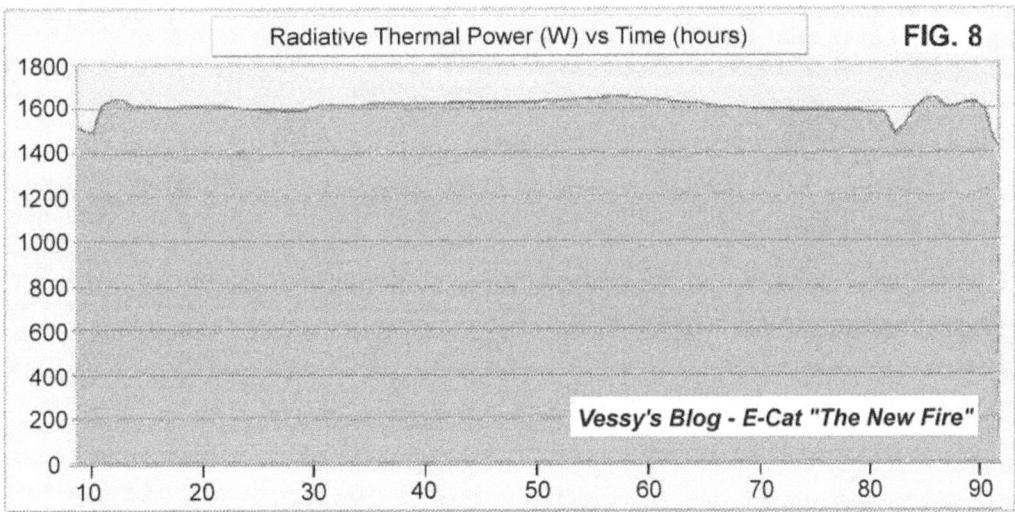

Fig. 2.4 - The almost constant radiative thermal power of the tested reactor, useful for estimating COP.

The **Coefficient of Performance** (COP) of the reactor device was obtained as the ratio between the total energy emitted by the device (radiated power + the power dispersed by convection) and the energy consumed by its three resistor coils. The resulting COP, with **many conservative assumptions**, was **5.6 +/- 0.8** (would be 4.5 taking into account only the radiative energy).

CHAPTER 3

THE HOT-CAT "TYPE III DESIGN"

The picture here below, published in a skinny version in the patent application filed by *Industrial Heat* on April 26, 2014, shows a layered tubular **reactor device** (Fig. 11 in the cited document), also represented in cross-sectional view (Fig. 12). It is **the third of three** different **embodiments** described in such patent application, so hereinafter I'll indicate it as "*E-Cat HT – III*".

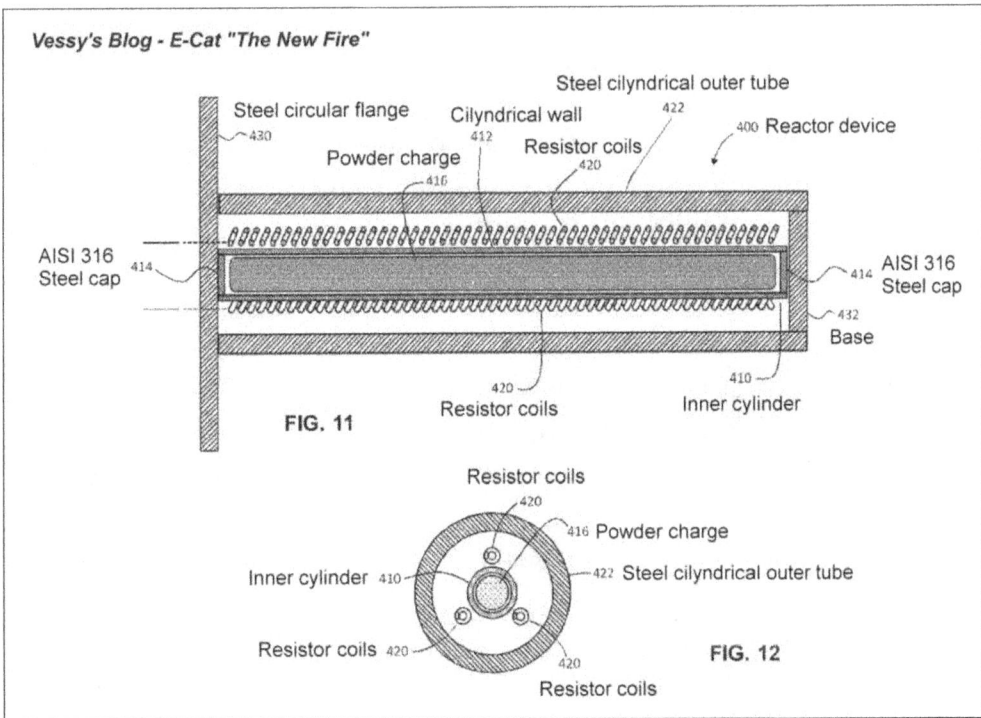

Fig. 3.1 - *Diagram of a reactor device E-Cat HT "Type III design" (from IH's patent, slightly modified).*

This reactor, with the powder charge widely and uniformly distributed along the central axis of the reactor, **was used in the third of the three tests**, or "experiments", described in the first *Third Party Report* (TPR-1).

Such experiment consisted in a 116-hours run **in self-sustained mode** of the device and was performed, successfully, on March 18-23, 2013 in Ferrara, Italy.

Document	TPR-1	TPR-1	TPR-1	TPR-2
Test date	Novemb. 2012	Decemb. 2012	March 2013	March 2014
Location	Ferrara (IT)	Ferrara (IT)	Ferrara (IT)	Lugano (CH)
Reactor type	E-Cat HT - I	E-Cat HT - II	E-Cat HT - III	E-Cat HT - IV
Temperature	793 °C	436 °C	302 °C	1260-1400 °C
Fuel	Ni, H, ?	Ni, H, ?	Ni, H, ?	Ni, H, Li, ?
Duration	Failed	96 hours	116 hours	32 days
Self-Sustained	No	No	Yes	No
COP	not available	5.6 +/- 0.8	2.9 +/- 0.3	3.2-3.6

Tab. 3.1 - All the tests described in the Third Party Reports released from the scientists Levi et al.

According to the dispersive description given in the cited patent and widely integrating the information contained on this issue in TPR-1, **the reactor device** used in this experiment **differed** from the so-called "Hot-Cat" reactors characterized by the earlier described Type I design and Type II design both **in structure and control systems**.

Externally, the reactor device (400) had a **steel cylindrical outer tube** (422) which was 9 centimeters in diameter and 33 centimeters in length, with at one hand a **steel circular flange** (430) 20 cm in diameter and 1 cm thick. An important purpose of such flange was to allow the reactor device to be supported while inserted in one of various **heat exchangers**.

The outer surface of the outer tube and one side of the flange were coated with **black paint**, different from that used for the second experiment. The black paint used was *Macota* enamel paint (produced by Macota srl, Italy), capable of withstanding temperatures **up to 800 degree** Celsius. The **distribution of the temperatures** along the device is not uniform, and the higher temperatures are reached in the central part of the cylindrical body.

Fig. 3.2 - The E-Cat HT "Type III design" before the Third Party test performed on March 18-23, 2013. Electrical power is fed through the two yellow wires, the third connection is a PT-100 sensor.

As in the previous reactors Type I and II, a **powder charge** (416) was contained within a smaller AISI 310 steel cylindrical inner tube (410). Such inner tube had a **cylindrical wall** (412) that was 3 cm in diameter and 33 cm in length. The inner tube was housed within the outer tube together with the **resistor coils** (420), and closed at longitudinal ends by two AISI 316 steel caps (414).

Electrical power was fed through the flange to power the resistor coils. The third connection was a **PT100 sensor** (418), used to give a **feedback temperature signal** to the control box in order to regulate the

ON/OFF cycle followed in this third experiment. The PT100 sensors are platinum resistance **thermometers** commonly used in industry.

The power supply used in this experiment was not a three-phase supply, but **single-phase supply**: that is, the Triac power supply used in the second experiment was replaced by a **controller circuit** having a three-phase power input and single-phase output, within a housing whose contents were not available for inspection, inasmuch as they are part of the **industrial trade secret**.

A significant difference between this Type III reactor device and the earlier described Type II reactor device lies in the **control system**, which now allows the reactor to work in **Self-Sustaining Mode** (SSM). That is, the reactor device can remain operative and active, while powered off, for **much longer periods** of time with respect to those during which power is switched on.

During this third experiment, after an **initial phase** lasting about two hours in which power fed to the resistor coils was gradually increased up to operating conditions, an ON/OFF phase was finally reached. In such ON/OFF phase, power to the resistor coils was **automatically regulated** by the temperature feedback signal from the PT100 sensor (see the resulting behavior here below).

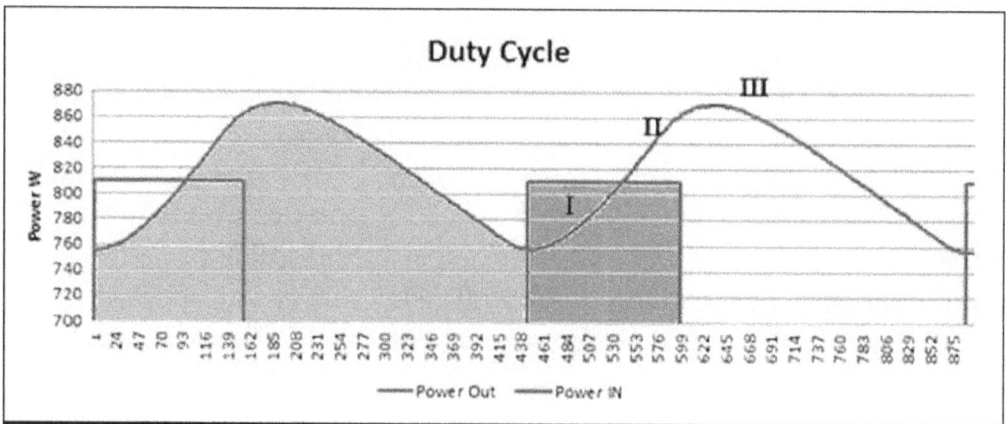

Fig. 3.3 - The first two periods of the ON/OFF cycle of the E-Cat reactor described in this article. It corresponds to Fig. 15 in the patent application and to Plot 8 in the TPR-1.

Upon initiation of the test, the **initial power input**, i.e. the instantaneous power absorbed, was about 120 W, gradually stepping up during the following two hours, until a value suitable for triggering the self-sustaining mode was reached. From then onwards, and for the following 114 hours of the tests, power input was no longer manually adjusted, and the **ON/OFF cycles** of the resistor coils followed one another at almost **constant time intervals**.

In the ON/OFF phase, the resistor coils were powered up and down by the controller circuit (402, see Fig. 13 of the patent) at observed regular intervals of about **2 minutes for the ON** state (equal to the 35% of the time, during which the instantaneous power absorbed was in the range of 910-930 W) and **4 minutes for the OFF** state (equal to the remaining 65% of the time, during which the temperature of the device continued to rise for a limited amount of time).

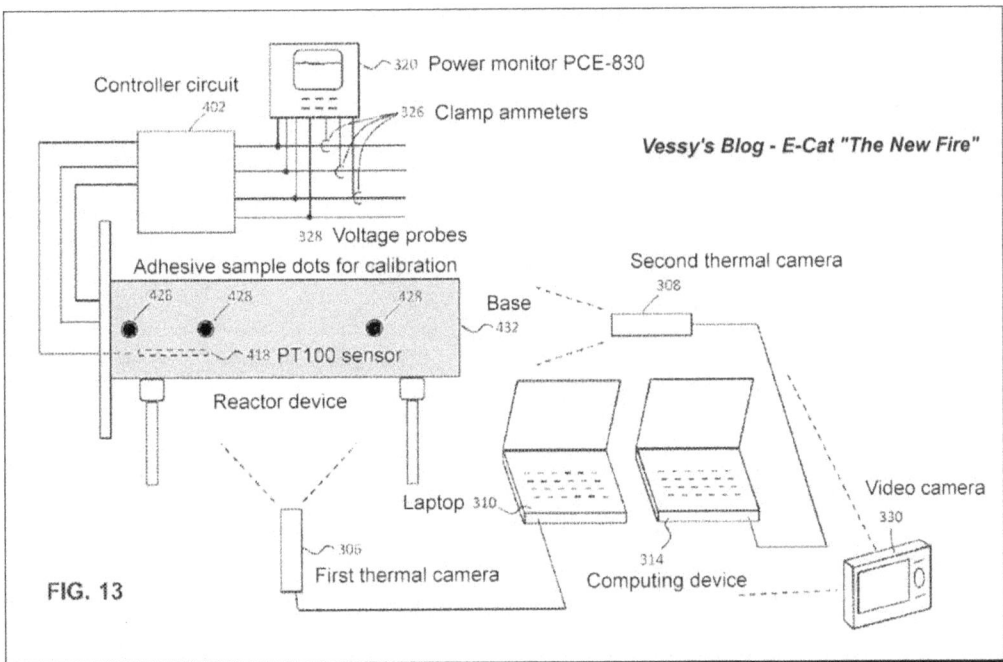

Fig. 3.4 - The experimental setup of the third test on a Hot-Cat reactor described in this article. You can use this image provided that you leave its attribution and a proper link.

A wide band-pass PCE-830 **power monitor** (320), in addition to providing voltage and current values for each phase, allows one to check both the **waveform** and its **spectral composition in harmonics** of the fundamental frequency (50 Hz). Voltage waveforms were confirmed as sinusoidal and symmetrical, and there were **no levels of DC** voltage. The instrument's stated measurement error is 2% within the 20th harmonics, and 5% from 21th to 50th.

As far as measurements of current are concerned, it was ascertained – by the third party team performing the test – that **no current** was present in the third phase and that, for the two other phases, the waveform harmonics spectrum, which appeared to be the one **normally associated with a Triac** regulator, was contained within the interval **measurable** by the instrument.

The **Coefficient of Performance** (COP) of the reactor in this third test was **2.9 +/- 0.3**. The reasons for the appreciable differences between the COP's values obtained in the 2nd and 3rd experiment (COP 5.6 and average temperature of 438 °C vs. COP 2.9 and average temperature of 302 °C) could depend on the **different quantities of powder** used and/or simply on the tendency of the COP, noticed even in the first experiment, to **increase with temperature**.

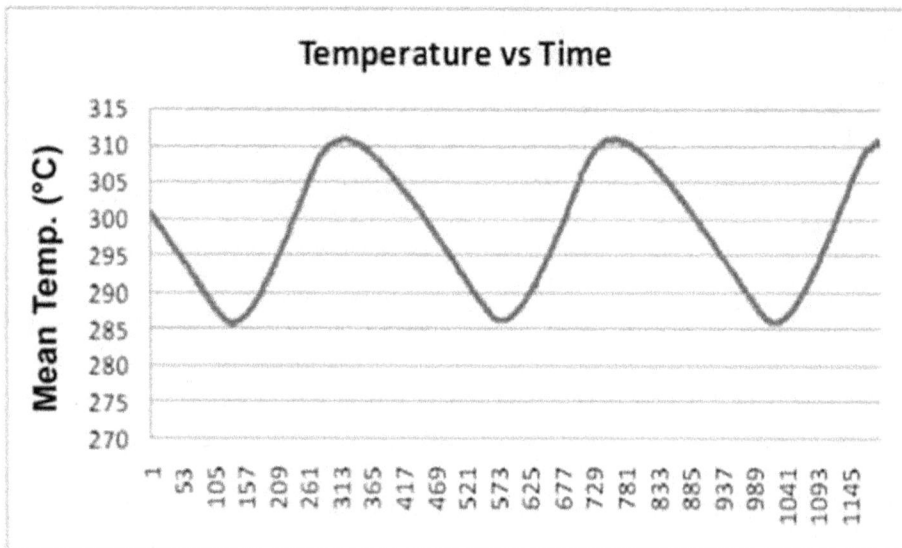

Fig. 3.5 - The average surface temperature of the E-Cat reactor in the test described in this article. It corresponds to Fig. 14 in the patent application and to Plot 3 in TPR-1.

This third test also included a **calibration** of the experimental setup **without the active charge** present in the E-Cat HT. In this case, no extra heat was generated beyond the expected heat from the electric input. The electrical power to this "unloaded" device was applied **continuously** as opposed to ON/OFF cycling, and stepped up gradually until a value of 910-920 W, including the power consumption of the **control box** (that was approximately 110-120 W).

Upon completion of this third test, the reactor device was opened, and **its parts weighed**. Before removal of the powder charges, the innermost cylinder sealed by caps was weighed (1537.6 g). Lastly, the inner powders were extracted and the cylinder was weighed once again (1522.9 g). The weight that may be assigned to the **powder charges** is therefore on the order of **0.3 grams**.

CHAPTER 4

THE HOT-CAT "TYPE IV DESIGN"

The image here below, a composition of two photos published in the 2nd Third Party Report (TPR-2) released on October 8, 2014, shows a "dog bone"-like **reactor device** used in a test performed in Barbengo (Lugano), in a laboratory placed at disposal by Officine Ghidoni SA. It is **the fourth of four** different known **embodiments** of a high-temperature E-Cat, or Hot-Cat, so hereinafter I'll indicate it as *"E-Cat HT – IV"*.

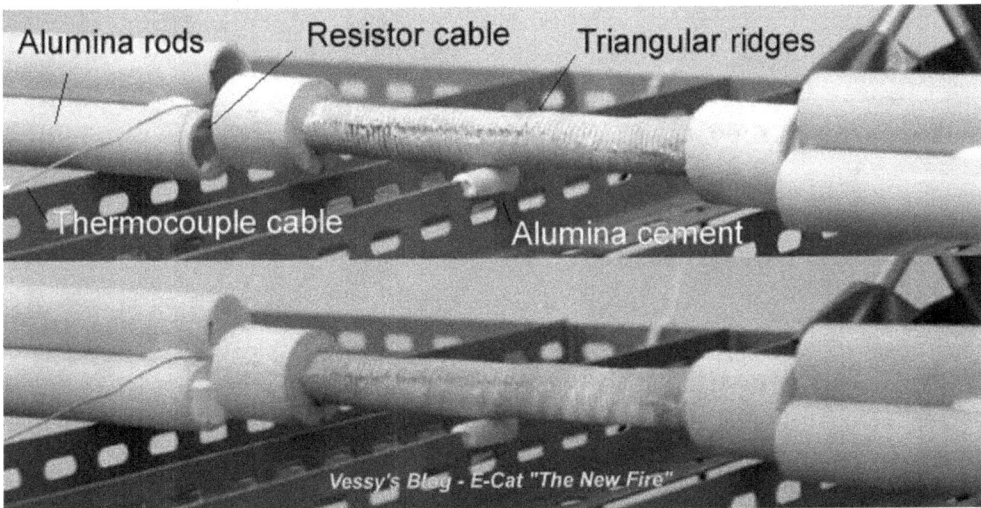

Fig. 4.1 - The "Type IV design" reactor, before the electrical ignition and during the test.

Data were collected during 32 days of running, from February 26 to March 29, 2014, a **much longer period** compared to the previous three tests described in the 1st Third Party Report (TPR-1), that I've illustrated in detail in previous chapters. The E-Cat reactor was tested by the **same collaboration** of scientists performing the tests described in TPR-1, in both the case with the financial support of Elforsk AB, Sweden.

Document	TPR-1	TPR-1	TPR-1	TPR-2
Test date	Novemb. 2012	Decemb. 2012	March 2013	March 2014
Location	Ferrara (IT)	Ferrara (IT)	Ferrara (IT)	Lugano (CH)
Reactor type	E-Cat HT - I	E-Cat HT - II	E-Cat HT - III	E-Cat HT - IV
Temperature	793 °C	436 °C	302 °C	1260-1400 °C
Fuel	Ni, H, ?	Ni, H, ?	Ni, H, ?	Ni, H, Li, ?
Duration	Failed	96 hours	116 hours	32 days
Self-Sustained	No	No	Yes	No
COP	not available	5.6 +/- 0.8	2.9 +/- 0.3	3.2-3.6

Tab. 4.1 - All the tests described in the Third Party Reports released from the scientists Levi et al.

A longer test was also motivated to investigate the **long-term stability** of the E-Cat operation, as well as running it at two different operational settings for comparison. The reactor operating point, indeed, was set at about **1260 °C** (corresponding to a measured electric power input of about 810 W) in the **first half of the run** (10 days), and at about **1400 °C** (corresponding to slightly above 900 W) in the second half (22 days).

It was chosen **not to induce** a self-sustained mode, i.e. the ON/OFF power input mode used in the third TPR-1 test. However, subsequent calculation by the authors of TPR-2 proved that increasing the input by roughly **100 watts** (from about 800 W to 900 W) had caused **an increase** of about **700 watts** in power emitted. The picture 4.2 shows the E-Cat net power production trend throughout the test.

The present E-Cat reactor is an improved version running **at much higher temperatures** then the earlier versions used in TPR-1's experiments, but avoiding at the same time **internal melting** – a previously fairly frequent occurrence – because in the course of the year following the previous tests the IP of E-Cat technology was transferred to **Industrial Heat**, USA, where the reactor was replicated and improved.

Thus, the reactor used in this occasion is outwardly quite different from the ones used in the tests held in the past years. Its external appearance is that of an **alumina cylinder**, 2 centimeters in diameter and 20 centimeters in length, ending on both sides with two cylindrical alumina blocks (4 cm in diameter, 4 cm in length), not detachable from the body of the reactor, which henceforth will be referred to as "**caps**".

Net Power Production

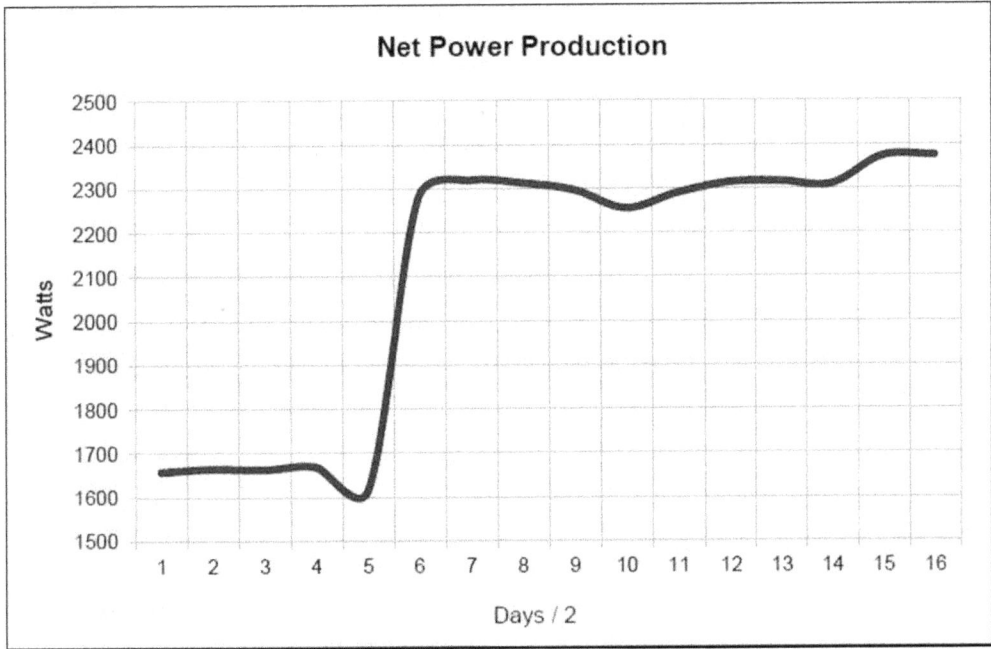

Fig. 4.2 - The E-Cat net power production trend throughout the test, given by the difference between the total watts produced by the reactor and the watts consumed by it. It shows how much emitted power is exclusively due to the E-Cat's internal reaction.

It is interesting to notice that, upon completion of the test, the authors of TPR-2 took a **sample of the material** constituting the reactor. Subsequently, Prof. Ennio Bonetti (who teaches Physics of Matter at the University of Bologna) subjected it to X-Ray spectroscopy. The results confirmed that it was indeed alumina, with a purity of at least 99%.

Whereas the surface of the caps is smooth, the outer surface of the body of the E-Cat is molded in **triangular ridges**, each 2.3 mm high and 3.2 mm wide at the base, covering the entire surface and designed to **improve the dissipation** surface for natural convective thermal exchange (cylinder diameter is calculated from the bases of the ridges).

The power supply cables run through the two caps. According to TPR-2, "three braided high-temperature grade **Inconel cables** exit from each of the two caps: these are the resistors wound in parallel non-overlapping coils inside the reactor". Moreover: "the reaction is primarily initiated by heat from resistor coils around the reactor tube". I've cited exactly the words used in the report because they are very important.

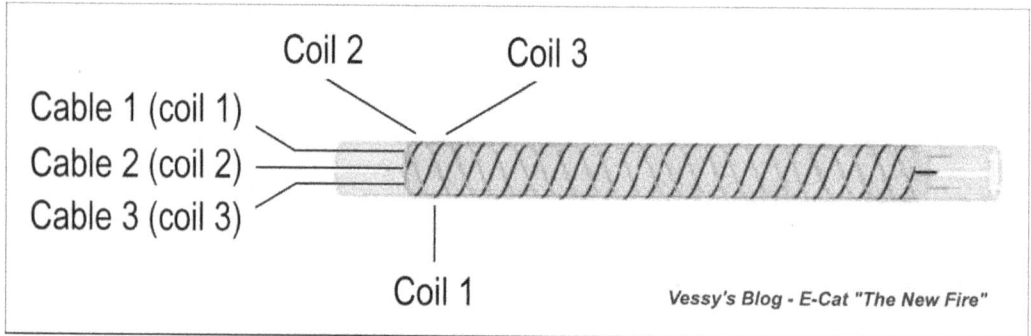

Fig. 4.3 - The three parallel non-overlapping resistor coils inside a "dog-bone"-like reactor.

As pointed out by the authors of the report, it is well known that some Inconel cables have a crystalline structure that **is modified** by temperature, and are capable of withstanding high currents only if they are operated **at the appropriate temperature**. If these conditions are not met, microscopic melt spots are liable to occur in the cables.

This generated some perplexity in the readers of the report. In response to questions about this issue, Andrea Rossi, some days after the release of TPR-2, admitted on his JoNP: "The coils of the reactor are made with a **proprietary alloy**, and the Inconel is only a **doped component** of it". It also said, in an open letter to Stephan Pomp, that "the resistances do not have a linear response to the temperature in the coil of the E-Cat".

A so-called "K-type" **thermocouple probe**, inserted into one of the caps, allows the control system to manage power supply to the resistors by measuring the **internal temperature** of the reactor. The thermocouple probe cable is inserted in an alumina cement cylinder, which acts as a **bushing** and perfectly fits the hole, about 4 mm in diameter.

The hole for the thermocouple probe is **also the only access point** for the fuel charge. When charging the reactor, the bushing is pulled out, and the charge is inserted. After the thermocouple probe has been lodged back in place, the bushing is sealed and secured with a mixture of **water and alumina powder cement**. To extract the charge, pliers are used to open the seal.

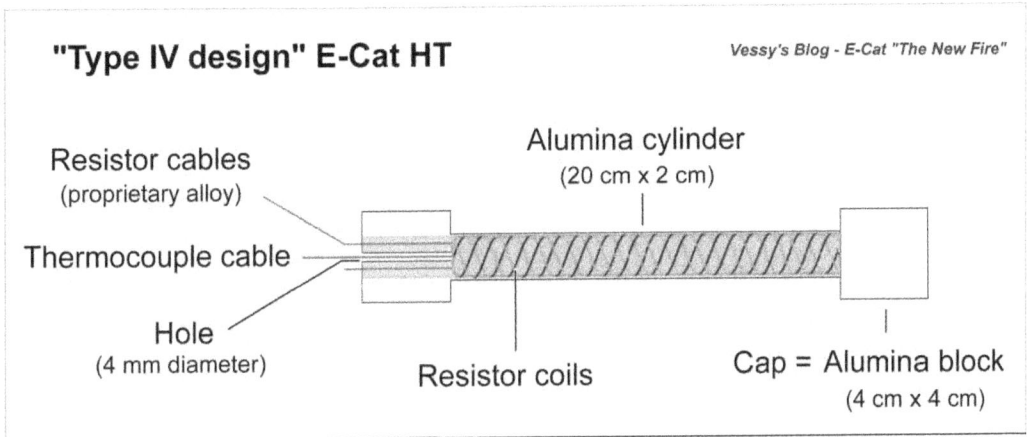

"Type IV design" E-Cat HT

Vessy's Blog - E-Cat "The New Fire"

Resistor cables
(proprietary alloy)

Thermocouple cable

Hole
(4 mm diameter)

Alumina cylinder
(20 cm x 2 cm)

Resistor coils

Cap = Alumina block
(4 cm x 4 cm)

Fig. 4.4 - Diagram, that I've derived from the above description, of a reactor device E-Cat HT "Type IV design".

The resistors and the copper cables of the three-phase power supply are connected, outside the caps, in the classic **"delta configuration"**, illustrated later. For 50 cm from the reactor, the power cables are contained in hollow alumina rods (three per side), 3 cm in diameter. The purpose of the rods is to **insulate thermally and electrically** the supply cables and protect the connections that run through them.

Both the reactor and the rods lie on a **metal frame**, the points of contact with the frame being thermally insulated with alumina cement. The whole frame lies on an insulating rubber mat on the floor. We found that the ridges made thermal contact with any thermocouple probe placed on the outer surface of the reactor **extremely critical**, making any direct temperature measurement with the required precision impossible.

The reactor is charged with a small amount of hydrogen-loaded nickel powder plus some **additives, mainly Lithium** (the others are a secret). The powder charge had been weighed before insertion in the reactor, resulting in about 1 g. Samples of both **fuel and ash** powders were taken and properly analyzed. In addition, the resistor coils are fed with some **specific electromagnetic pulses**, also covered by industrial trade secret.

The E-Cat's control apparatus consists of a three-phase **Triac power regulator**, driven by a programmable microcontroller; its maximum nominal power consumption is 360 W. The regulator is driven by a **potentiometer** used to set the "operating point" (i.e. the current through the resistor coils, normally 40-50 Amps), and by the **temperature read** by the reactor's thermocouple.

The input power was carefully monitored with appropriate instruments: two **PCE-830** for electric power measurements, and **three digital multimeters** to verify that no DC components were present in the power supply. The output power was determined by measuring, with two high-resolution **thermal imaging cameras** *Optris PI 160 Thermal Imagers*, the emitted radiation as well as calculating the heat dissipation from convection.

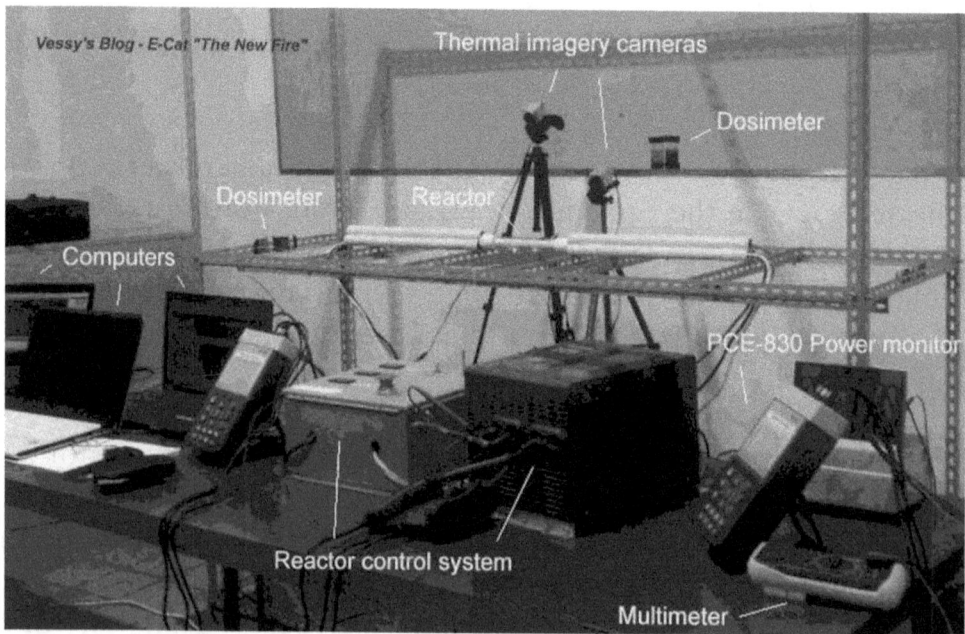

Fig. 4.5 - Photo of the experimental setup used for the measurements (text added).

All the instruments used during the test are **property of the authors** of TPR-2, and were calibrated by their respective manufacturers, but a further check was made to ensure that were not yielding anomalous readings. Throughout the test, all the above instruments were connected to the **same computer**, wherein all the acquired data were saved. For both the PCEs and the IR cameras, **data acquisition frequency** was set at 0.5 Hz.

The two PCEs are located **one upstream and one downstream** – as shown in the figure below – from the control instruments, the already mentioned Triac three-phase power regulator driven by a potentiometer and by the temperature read by the K-probe. Note that, in the picture, the three cables running from the control system to C are termed C1, whereas the six cables running from C to the reactor are termed C2.

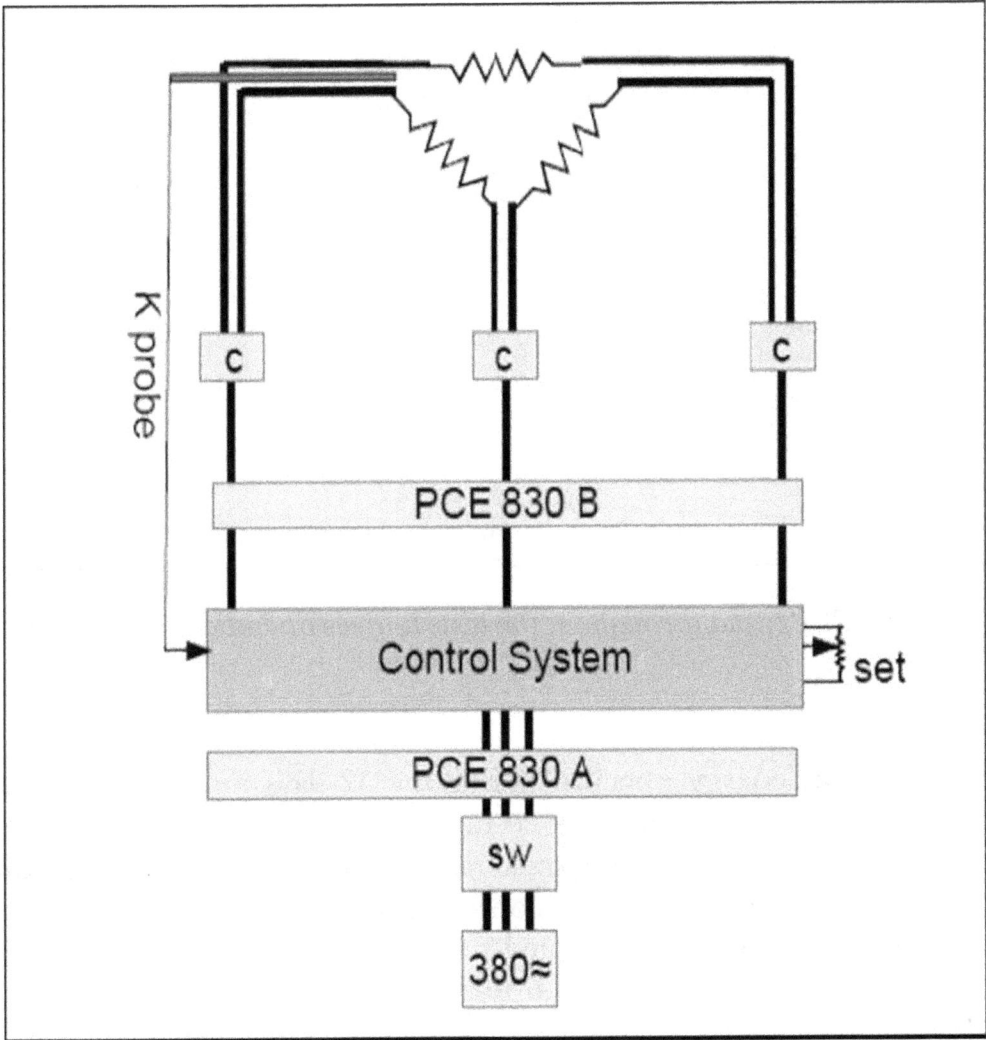

Fig. 4.6 - The wiring diagram of the test to a "Type IV design" Hot-Cat, as described in TPR-2. The resistors are connected in the so-called "delta configuration" (SW = Switch, C = Connection Box).

The **COP**, calculated as the ratio of the sum of the mean power emitted – by radiation and convection by both the E-Cat and the rods – to mean power consumption of the reactor (minus watts dissipated by the cables through Joule heating), resulted **3.1 +/- 8%**, corresponding to a net production of 1658 W in the first half of the run (10 days), and **3.6** in the second half (22 days).

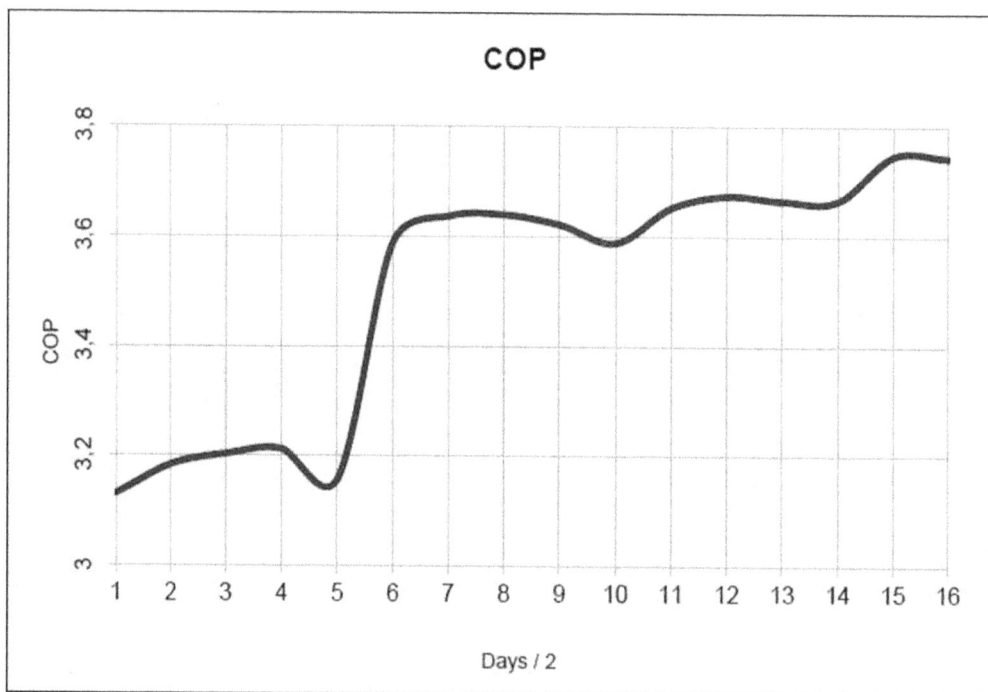

Fig. 4.7 - COP trend throughout the test. It gives an indication of the E-Cat's performance.

The **total net energy** obtained during the 32 days run was about **1,6 MWh**. This amount of energy is far more that can be obtained from any known chemical sources in the small reactor volume. Indeed, the authors of TPR-2 estimated – in a very conservative way – the **power density** and the **energy density** associated to the E-Cat's fuel, which resulted, respectively, **2.1·10⁶ W/kg** and **1.6·10⁹ Wh/kg**.

CHAPTER 5

A POSSIBLE ALTERNATIVE DESIGN

In this chapter, I will describe a new design for a Hot-Cat, but this time the description does not refer to a reactor designed by Andrea Rossi or Industrial Heat, but to an **independent "replication"** of the Hot-Cat, made by a 70 years old Russian physicist with a solid scientific curriculum – Alexander G. **Parkhomov** – which seems to have obtained results similar to those described in my previous chapter for Type IV reactor.

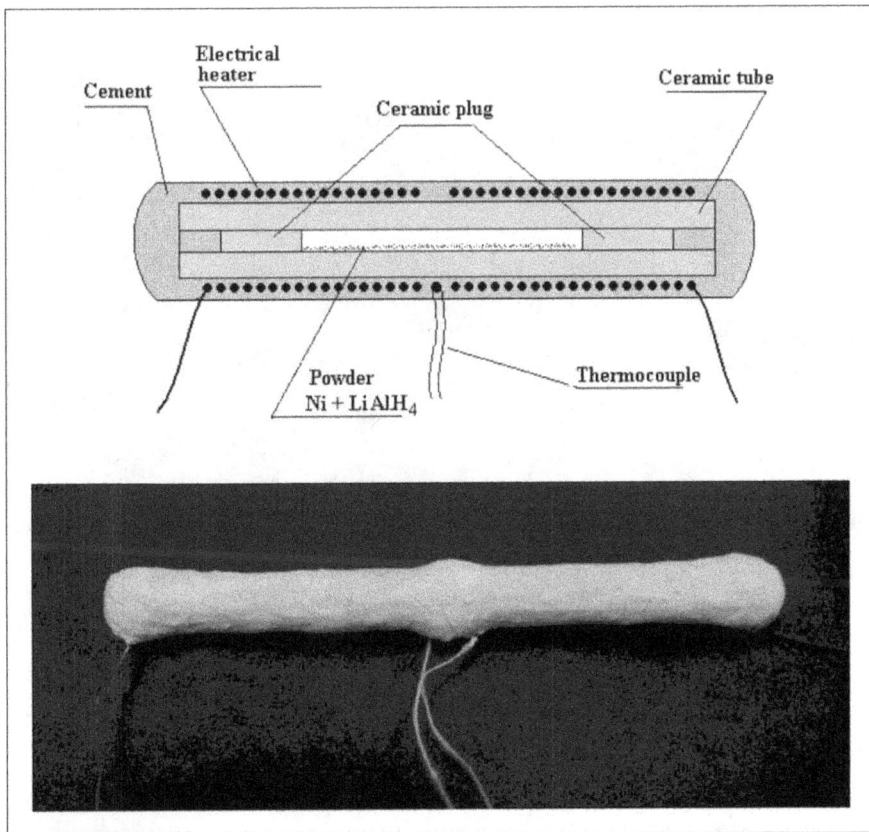

Fig. 5.1 - Design of the Parkhomov's reactor (above) and the reactor prepared for the experiment (below).

Alexander Georgievich Parkhomov, a disciple and colleague of Nobel Prize winner Andrei Sakharov, was born on January 31, 1945. In 1968 he finished his studies at the faculty of "Experimental and Theoretical Physics", at the **Moscow Engineering-Physical Institute**. In 1975, he earned a PhD at the faculty of "Physics of Radiation" of the same institute. In 1979-81 he worked as a senior scientific employee at the Vladimir Polytechnical Institute, and from 1987 to 1993 headed a **research group** at the Aviation Institute in Moscow. He's author of about 100 scientific papers, of which 45 with co-authors.

Parkhomov described his reactor, almost a replication of Rossi's Hot-Cat – namely of the Type IV design described in TPR-2 – in the **slides of a presentation** published online, prepared for his seminar entitled "Cold Fusion and Light Balls", held at the People's Friendship University of Russia on December 25, 2014. He appears to have reached a **COP = 2.58** at a temperature $T = 1290$ °C, so its reactor deserves great attention.

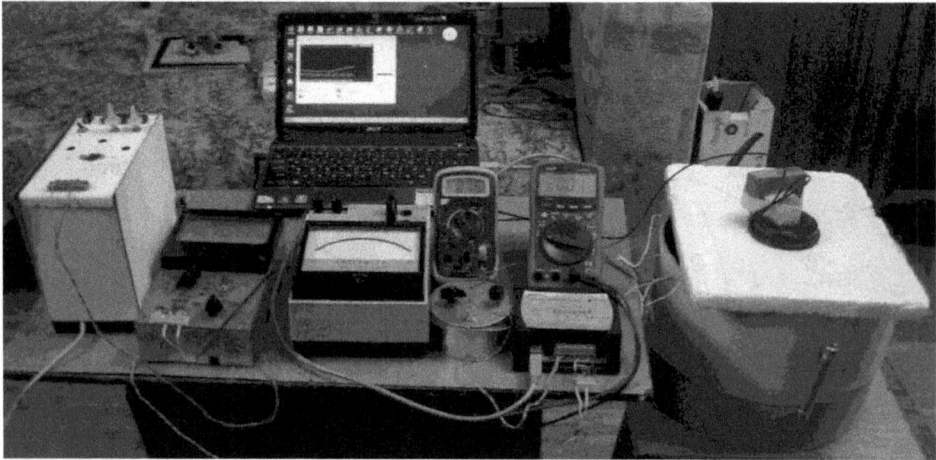

Fig. 5.2 - The simple setup used in the Parkhomov's experiment.

The reactor built by Parkhomov in his home laboratory **differs** from the Rossi's "Dog Bone" in many aspects, of which the main ones are: (1) **geometry**, which is similar to the "Type IV design" but not identical; (2) the fact of not to use a three-phase electric power, but a normal **single-phase** power; (3) perhaps the **lack of one or more** chemical components of the catalyst used by Rossi, still secret for reasons of Intellectual Property.

For these reasons, we can speak of a sort of "Type V" design distinct from the previously described "Type IV" design of Rossi's reactor. In addition, Parkhomov's reactor uses a **very simple experimental setup**, especially as regards the measure of excess heat. Instead, little is known on how atmosphere was removed from the reactor. And there are no pictures of the fuel after the test was completed to see if it was sintered.

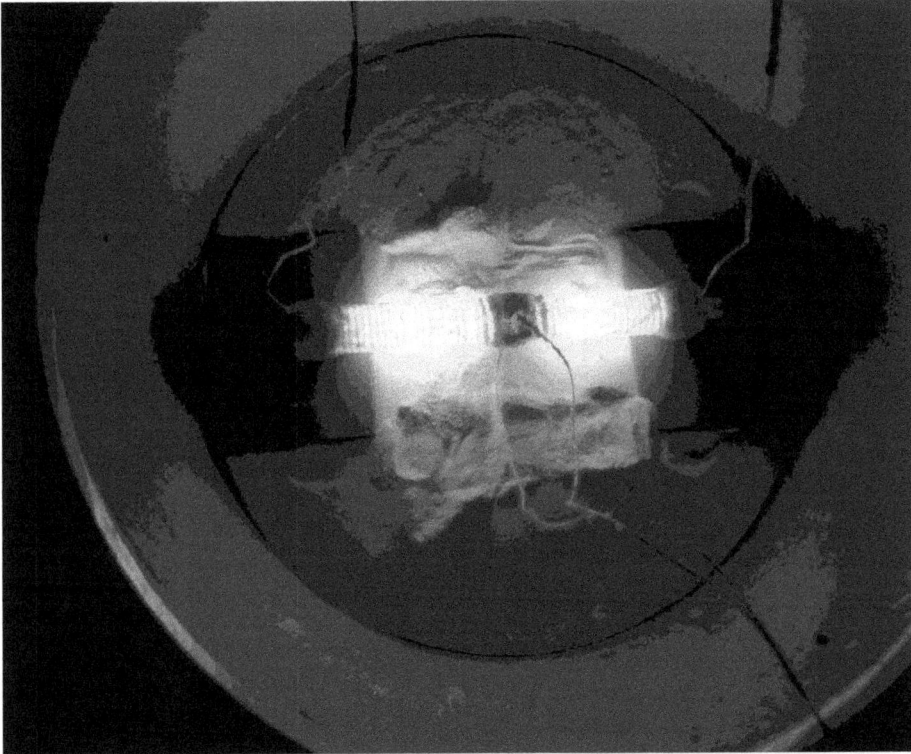

Fig. 5.3 - The reactor during operation. The covers of the thermal insulation and of the vessel have been removed.

We can see in the photo 5.2 the essential measurement equipment used by Parkhomov. From right to left of the image: **power supply** for heaters, **Geiger counter** display, **ammeter**, thermocouple amplifier, reactor temperature display, computer data logger, **digital voltmeter**. At the right side: Reactor in the calorimeter. On the top: Geiger detector. On the side surface: **radiation dosimeter**. A laptop computer is used for data logging of the reactor temperature and Geiger counter.

For measuring the consumed electric energy the "Mercury 201" electrical counter was used which allows the transfer of the information to the computer, also from the voltmeter and ammeter. During the first experiments the electric supply for heating the reactor was taken directly from the mains using **thyristors**.

Later experiments used a **changing transformer winding**. Both manual and automatic switching was used by the temperature controller. This allows us to provide continuous operation of the reactor at the given temperatures, improving the stability of functioning of the reactor. You can see the **circuit diagram** in the picture below.

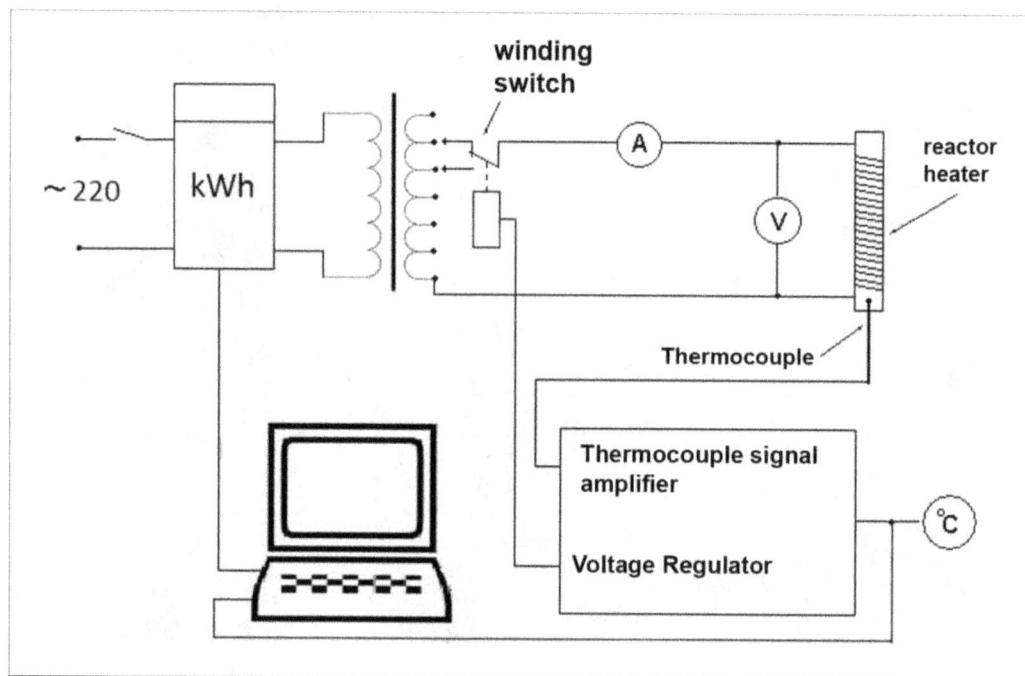

Fig. 5.4 - Circuit diagram of the power supply and control system used by Parkhomov.

Measurement technique of the liberated heat used by Parkhomov to check the performance of his reactor is simple: it is based on the amount of **vaporized water**, because that based on IR thermal images used in the Third Party Reports was too complex. The measurement method adopted by the experienced Russian physicist had been elaborated and experimentally verified multiple times by his colleague Yuri N. Bazutov.

How the **calorimetric measurement** is made? You can see in the picture 5.5 the simple calorimeter used by Parkhomov. The reactor is enclosed in a **metallic vessel**, which has a heavy cover. Thermal insulation is made of foam and on this cover a Geiger counter is placed. This vessel is **immersed in water** inside the outer vessel.

When the water boiling begins, part of it boils away in the form of **steam**. The released heat quantity is estimated by measuring the water quantity **before and after** the experiment. By measuring the decrease of water, and from the known **heat of vaporization** (2260 joules/kg), it is easy to calculate the generated heat. The correction for the heat lost through the isolation can be estimated from the heat cooling rate after reactor shutdown.

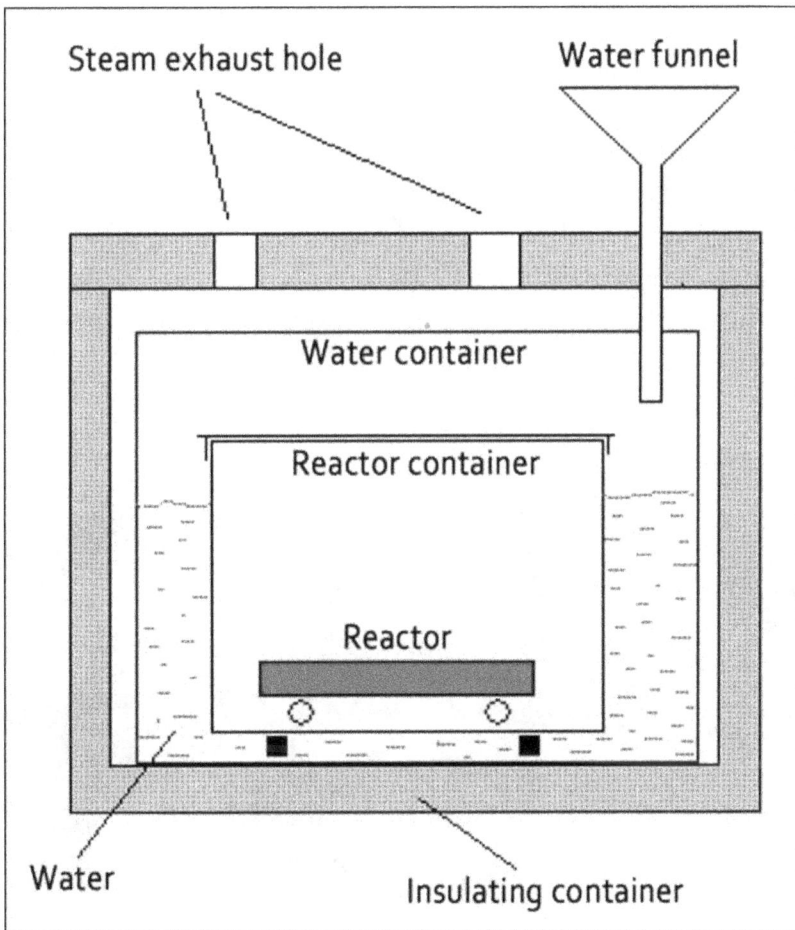

Fig. 5.5 - Design of the simple calorimeter used by Parkhomov.

Therefore, the measurement of the amount of water vaporized in the control and active runs is very simple. As Parkhomov explains: "During the experiment, after boiling of water began I kept **invariable** the top level of this water by **gradual addition** of fresh water. The mass of the evaporated water was considered to be equal to the mass of the added water".

The expert of LENR Michael McKubre commented positively this type of calorimetry: "The method has been employed accurately for well over 100 years. With simple precautions, it should be **accurate within a few percent** over a wide range of powers and reactor temperatures. One must be concerned to interrogate the heat that leaves the calorimeter by means other than as steam escaping at ambient pressure, that water does not leave the vessel in the liquid phase as splattered droplets or mist (fog), and to **accurately measure the water mass loss** (or its rate) to determine output power".

As revealed later by Parkhomov himself in a private communication to Frank Acland of *E-cat World*, measurements with a reactor **not containing fuel** and an electrical heater at the power up to 1000 W were taken. The quantity of consumed electric power after boiling of water, and the amount of heat necessary for heating and evaporation added for preservation of initial level, **coincided within 10%**.

Fig. 5.6 - A young Alexander Parkhomov in his home laboratory.

The reactor consists of a ceramic tube **made of alumina** (Al_2O_3). Such cylinder has a length of 120 mm, an external diameter of 10 mm and an internal diameter of 5 mm. An electric heater is wounded around the internal tube thick 2.5 mm. The heater is made of a heat-resistant alloy **"nichrom"**. The wire was coiled directly on the corundum tube with intervals of 0.5 mm between rounds, and then was covered with heat-resistant **cement**.

As Dr. Parkhomov responded to an inquiry regarding the construction of the reactor he made: "The search of cement, which maintains high temperature, was **the most complex problem**, which should be faced preparing the experiment. You need not only the right chemical structure, but also a **process engineering** – for application of the cement – including some stages. The creation of the reactor lasted 3 days".

Here below you can see the sizes analysis of the nickel powder used in the **fuel mix,** as provided by Parkhomov in reply to a specific question made by Frank Acland. We see that the **mean size of the grains is 12.85 μm,** and that less than 10% of the powder is smaller than 3.150 μm. Probably, Parkhomov used this **wide range distribution** to increase the likelihood that some Rossi-like effect was found.

Volume Statistics (Arithmetic) N⁰ 4				
Calculations from 0.017 μm to 2000 μm				
Volume:	100%			
Mean:	12.85 μm	S.D.:		11.33 μm
Median:	8.724 μm	Variance:		128.3 μm²
Mean/Median ratio:	1.473	C.V.:		88.1%
Mode:	7.083 μm	Skewness:		1.587 Right skewed
Specific Surf. Area:	1007 cm²/g	Kurtosis:		1.953 Leptokurtic
d_{10}: 3.150 μm		d_{50}: 8.724 μm		d_{90}: 31.28 μm
<10%	<25%	<50%	<75%	<90%
3.150 μm	4.857 μm	8.724 μm	16.55 μm	31.28 μm

Fig. 5.7 - The sizes analysis of the nickel powder used by Parkhomov.

The outer surface of the cylinder is in contact with a thermocouple, which is placed in the central part of the tube, as shown in the design picture of the reactor, illustrating also the cables of heater and thermo-couple. The **ends of the tube** are sealed with high temperature resistant ceramic cement. The whole **surface of the cylinder** is covered by the same cement (likely, alumina, already used for the ceramic tube).

The Parkhomov's "Hot-Cat" is loaded with a mixture of nickel and of so-called "lithium aluminum hydride" (LiAlH$_4$). More exactly, inside the tube there is **1 gram of a powder of pure Nickel + 10% LiAlH$_4$** (ten percent by weight). The quantity of lithium aluminum hydride was not chosen at random, but according to a precise quantitative evaluation.

When lithium aluminum hydride is heated, there is a decomposition of such chemical compound and hydrogen is progressively released as a gas, so its pressure in the reactor increases. We can see, with a simple calculation, that **1 g of lithium aluminum hydride** allocates **0.105 g of hydrogen** or, going from weight to volume, delivers **1.17 l** of hydrogen (at normal air pressure and temperature).

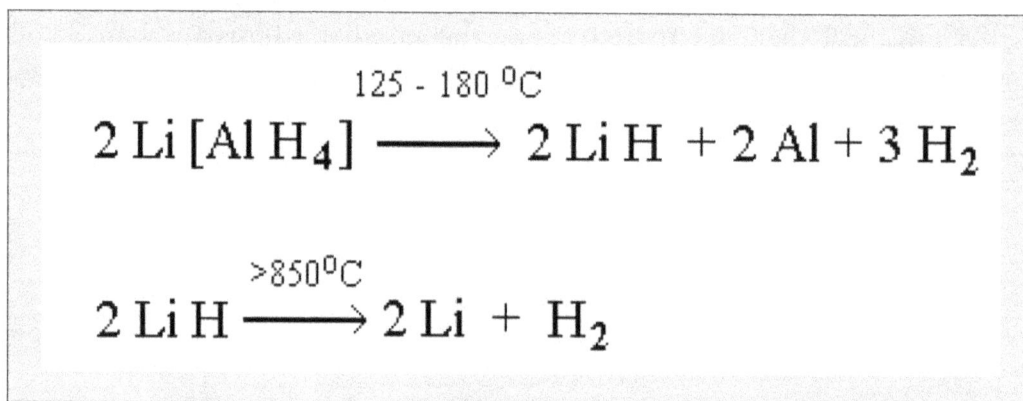

$$125 - 180\ ^{0}C$$
$$2\,Li\,[Al\,H_4] \longrightarrow 2\,Li\,H + 2\,Al + 3\,H_2$$

$$>850^{0}C$$
$$2\,Li\,H \longrightarrow 2\,Li + H_2$$

Fig. 5.8 - Decomposition of the lithium aluminum hydride releasing hydrogen.

Assuming that the interior of the reactor is a channel with a diameter of 4 mm, the **cavity volume** is approximately **2 ml**. From the previous calculation, we know that 100 mg of lithium aluminum hydride deliver about 100 ml of hydrogen (under normal pressure and temperature conditions). So, if 100 ml are compressed up to 2 ml, at typical air temperatures the pressure rises to **50 atmospheres**.

At the working temperature of Rossi's reactor, the nickel mixes with the liquefied aluminum and a gas environment of lithium and hydrogen appears. When the temperature exceeds 1,000 °C, residual air reacting with the hydrogen, lithium and aluminum, under a pressure that may reach **over 100 atmospheres**, makes a small quantity of nitrogen, ammonia, nitric oxide and oxides of lithium and aluminum.

The heating power supplied to the heater inside the reactor – through a standard 50 Hz AC with no other frequency stimulation, wave chopping or magnetic field – has been varied **stepwise from 25 W to 500 W**, and **after 4 hours** the external temperature of the reactor reached 1,000 °C. This is reflected in a stepwise increase shown in the picture below.

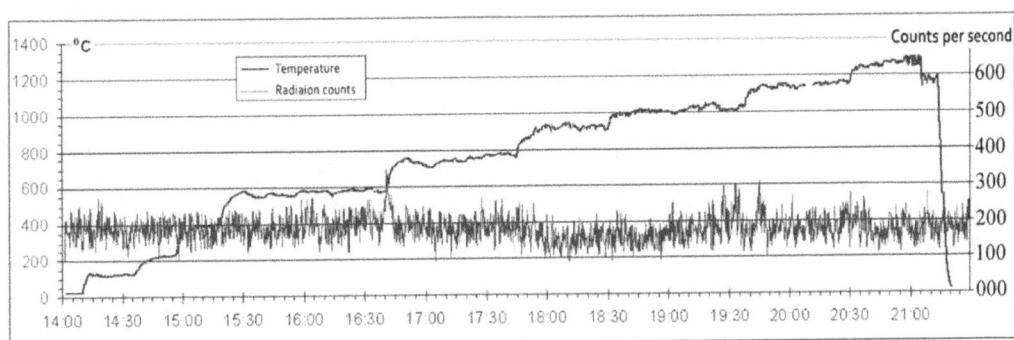

Fig. 5.9 - *Temperature of reactor and counts from a Geiger counter during the experiment.*

The same diagram shows the count rate (in counts/sec) provided by a **Geiger-Muller counter** – likely homebuilt and connected to a PC through a data logger – based on the Russian high sensitivity pancake probe SI-8B. This type of Geiger probe responds to **alpha, beta, gamma and X-rays**. It can be seen that, throughout the entire heating phase, the count rate is **not very different** from the background level (around 2 pulses/sec).

A small negligible increase in the background radiation is observed **sporadically**. Indeed, some spikes reaching a value of 3 counts/sec are noticeable at temperatures around 600 °C and 1000 °C. Further studies are thus needed to show if it was only an occasional occurrence or a sort of pattern. However, a **pocket dosimeter type DK-02** did not find a radiation in the limits of the errors of measurement (5 mR).

For neutron detection, Parkhomov used a **foil of Indium** immersed in the water of the calorimeter, however there was no observable activation of the foil during his experiments (so neutron flux density does not exceed 0.2 neutrons/cm^2). The **activity** of the Indium was measured using two Geiger counters. The impulses of the counters were recorded by a specialized computer. The same computer records the impulses from the Geiger tubes put above and below the dosimeter film and the metered electricity consumed.

The figure below shows in detail the temperature change occurring at the electrical heating power of 300, 400 and 500 watts. You can see that, at a constant heating power, there is a **gradual increase in temperature**, particularly strong in the last part of the run at 500 W of constant heating power. At the end of the part with the highest temperature visible in this chart, some temperature **oscillations** begin.

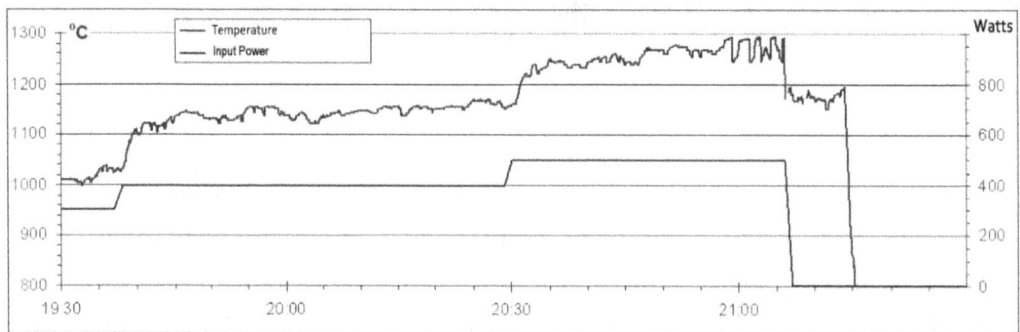

Fig. 5.10 - Changes in the heating process: zoom in the area of high temperatures.

The entire run of the Parkhomov's reactor ends due to the interruption of the electrical heating, as a result of a **burning of the heater**. After that, **for 8 minutes** the temperature stays at almost 1,200 °C, and then begins to fall sharply. This indicates that in the reactor, at this time, heat is produced at a high level **without** any electrical heating.

Thus, for this time duration (8 min) a heat **at kilowatt level** is produced inside of the reactor without electrical heating. So, the plot shows that the reactor is capable to generate a significant heat power that is greater than the electrical heating. According to the Parkhomov's report, during the total working time (90 minutes), the **excess heat energy** produced by the reactor has been about 3 MJ or 0.83 kWh.

A table published in the presentation prepared by the Russian physicist shows the extracted heat and the thermal efficiency calculations made for the **three "modes" of operation** corresponding to a temperature of about 1000 °C, 1150 °C and about 1200 to 1300 °C. You can see that, at temperatures of 1150 °C and in the range of 1200-1300 °C, the heat produced by the reactor is **much greater** than the consumed energy.

Average temperature of a "mode"	°C	970	1150	1290
Duration of a "mode"	min	38	50	40
Electring heating power	W	300	394	498
Electrical energy consumption	J	684000	1182000	1195200
Mass of vaporized water	kg	0,2	0,8	1,2
Energy spent for vaporization	J	452000	1808000	2712000
Heat leakage rate through thermal insul.	W	155	155	155
Cumulative heat loss through thermal ins.	J	353400	465000	372000
Cumulative net energy	J	805400	2273000	3084000
Output/input energy ratio (COP)	COP	1.18	1.92	2.58

Tab. 5.1 - Estimate of released energy as heat. Calculations are made for three modes of operation with a temperature of about 1000 °C, about 1150 °C and 1200-1300 °C.

The COP of the Parkhomov's reactor, as shown from the above table, at 500 W of power input (average effective power input = 1290 W) resulted equal to 2.58. As, from what Parkhomov said, the energy output in the **dummy run was within 10%** of the calculated baseline based on water chemistry (corresponding to a COP of 1.1 instead of COP = 1.0), you would need more than 200% error to get a COP of 2.5.

So, the calculated COP is **significantly above** the error margin and there are no doubt about the generated excess heat, and the Parkhomov's experiment has shown that this device actually produces more energy than it consumes. It represents the **first confirmation** of the main results obtained in the Lugano test described in TPR-2. This is also the first fully independent "replication" by "somebody skilled in the art".

We've described the experiment carried out by Parkhomov on December 20, 2014. On January 2015, he has performed **other experiments** on a reactor with fuel and on a dummy, i.e. heating the reactor without fuel. The table shows the results obtained in several experiments. In experiments with reactor models **having no fuel** as well as with reactors with fuel at a temperature below 1000°C, the COP is **close to 1**.

Reactor with Fuel

Date	Temp	Duration	Input	Output	COP
	°C	Min	Watts	Watts	
20.12.2014	970	38	301	297	0.99
20.12.2014	1150	50	395	758	1.92
20.12.2014	1290	40	499	1365	2.74
04.01.2015	940	131	304	305	1.00
04.01.2015	1020	75	377	407	1.08
10.01.2015	1080	73	161	284	1.77
18.01.2015	800	90	308	293	0.95
18.01.2015	1080	38	78	135	1.73

Electric Heating Without Fuel

Date	Temp	Duration	Input	Output	COP
	°C	Min	Watts	Watts	
02.01.2015	210	56	211	227	1.07
02.01.2015	470	00	400	414	0.95
02.01.2015	1050	16	928	1035	1.12
21.01.2015	1000	69	297	296	1.00
21.01.2015	1080	43	306	297	0.97
28.01.2015	900	65	95,5	105	1.08
28.01.2015	1100	66	116	116	1.00
28.01.2015	1200	50	151	147	0.97

Tab. 5.2 - The results obtained by Parkhomov in his several replication experiments.

The **COP vs. temperature relationship** emerging from all these experiments has been shown by the reader "Sanjeev" in a chart (see Fig. 5.11) posted on the blog *E-Cat World*. I've added to the chart the points corresponding to the Lugano test and to the second and third test described in TPR-1. We can see from the extrapolated curve that probably **no catalyst was used in the Lugano test**, whereas it was used in the other tests on the Hot-Cat.

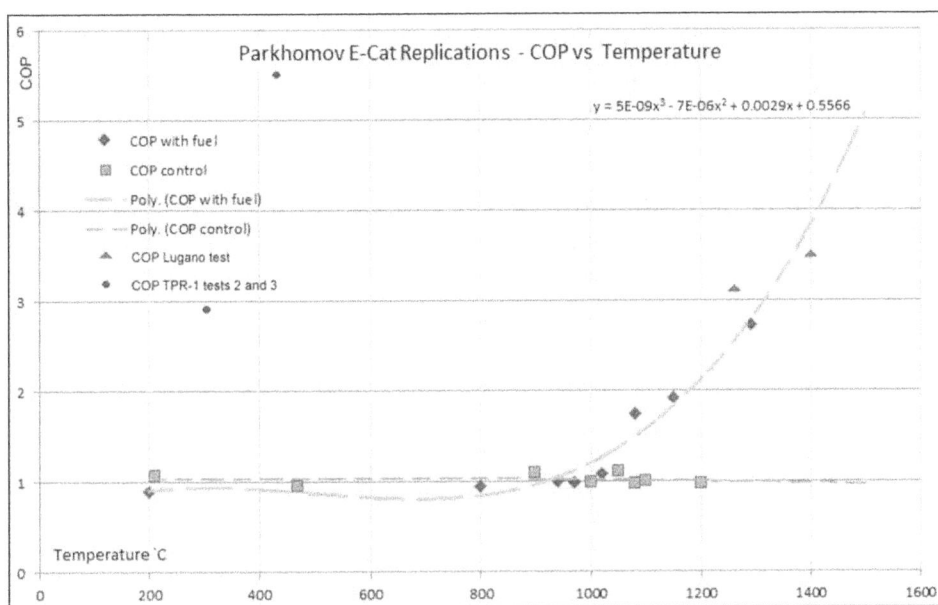

Fig. 5.11 - *The many Parkhomov's tests vs. both Lugano test and TPR-1's test 2 and 3. We can see that probably no catalyst was present in Lugano test.*

Sometimes, the reactor made of alumina powder was poured into a **metal envelope** to provide thermal insulation. This allows a 2-3 times reduction in the power necessary to heat the reactor; however, the operation in this regime is **less stable** than in case of the "naked" reactor. According to Parkhomov, many times in his experiments an uncontrolled **local overheating** resulted in destruction of the reactor.

Fig. 5.12 - *Some reactors after the experiments carried out by Parkhomov.*

In another experiment, the reactor was covered with a **further thermal insulation** of alumina powder instead of the metal envelope, then it worked for 38 minutes at a temperature near to 1080 °C, but when Parkhomov tried to increase the temperature the heater burned out. Thus, the main problem is **short-term operation** of the reactors, associated with the **destruction** caused by local overheating.

CHAPTER 6

THE SECRET INTERIOR OF A REACTOR

This chapter analyses, in the light of the information about the Hot-Cat we have today, the **official report** on the test performed on July 16, 2012, in Bologna by the Italian nuclear engineer **Fabio Penon** for the product certification at that time in course on the Hot-Cat. Indeed, it is interesting to compare this reactor with the Hot-Cats Type I-IV that we know in detail thanks to the TPRs and to the 2nd Patent application.

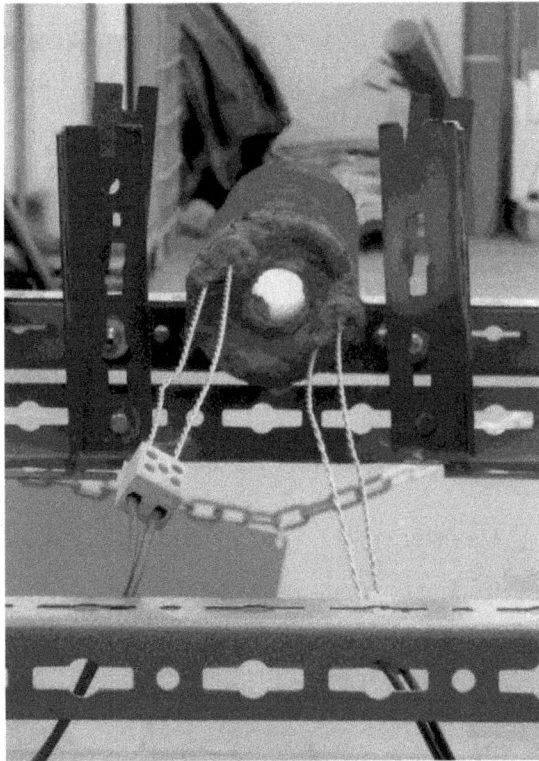

Fig. 6.1 - The Hot-Cat glowing white hot from its central hole during the Penon's test.

The Penon's test was not only a test focused on the certification process of Hot-Cat technology, but was also **preliminary** to the test described in the more detailed and comprehensive Third Party Report (TPR-1), whose first test occurred four months later (see Table 6.1 of all the official tests performed on Hot-Cats). For this reason, Penon's test was made unofficially **with 6 professors** of two Universities.

Year	Month	Reactor tested	Temperature	Duration
2012	July 16	Hot-Cat for Certification (Eng. Penon)	801 °C	6.5 hours
2012	November	Hot-Cat Type I design for TPR-1	793 °C	Failed
2012	December	Hot-Cat Type II design for TPR-1	436 °C	96 hours
2013	March	Hot-Cat Type III design for TPR-1	302 °C	116 hours
2014	March	Hot-Cat Type IV design for TPR-2	1260-1400 °C	32 days

Tab. 6.1 - A table showing all the official tests performed on Hot-Cat reactors.

The reactor tested is a single module consisting of an assembly of **four components** (an external cylinder, an internal cylinder, two heating resistors, four cables), plus the **active charge** (Nickel and a secret catalyst) and a **tablet** which acts as a hydrogen reserve.

The first component of the reactor device is an **external cylinder** made of **AISI 316** paint-coated stainless steel: its length was 33 cm, Ø 8.559 cm, weight 1272.7 g, as measured before assembly. The cylinder was subsequently painted externally and internally with **black coating** – heat resistant up to 1200 °C – to increase its emissivity.

Fig. 6.2 - The reactor dismantled after the test in a photo that shows how it's made.

The second component is the **inner cylinder**, made of the same stainless steel alloy. Here are its specifics: length 33 cm, Ø 3.385 cm, weight 705 g, as measured by the present examiners before assembly. The cylinder was subsequently painted on both outer and inner surfaces with black coating, heat resistant **up to 1200 °C**.

The third component of the reactor is a set of **two 230 V heating resistors**, total weight 2292.8 g. Their resistance values measured at the start of the test are 11.8 and 12.2 Ω, respectively. In parallel, the recorded value was 6 Ω. The two resistors were **powered in parallel**. Their maximum rated power can be calculated and is 8.069 kW.

Fig. 6.3 - The two 230 V heating resistors that we find inside the reactor, parallel to the tube.

Lastly, as fourth component there are **four ceramic cable fittings** for the resistors, each weighing 6.1 g, for a total of 24.4 g. These cables are connected to a resistor power supply control box provided with integrating wattmeter panel, manually controlled. A **voltmeter** and a **clamp ammeter** are connected to resistors downstream from the control box, so as to monitor power data independently from the panel meter.

Where is the powder charge? In the report there is **no mention** of where it is placed inside the reactor. However, according to an engineer present at the test (and likely author of the report), who wrote the following comment on an Italian blog with the **alias 'Cures'**: "The internal space between the two cylinders contains the heating resistor and the **reaction chamber** with the active material".

The reactor was sealed with **Saratoga TE7 1200**, a 1200 °C smelter-grade putty sealant that we can see very well in the many photos taken after completion of the test, when, after a long period of cooling, the E-Cat cylinder **was dismantled** to verify that the components were still the same, and to weigh the Saratoga sealant putty. This Saratoga mastic has been developed for blast furnaces, so there is **no need** for sealing pressure.

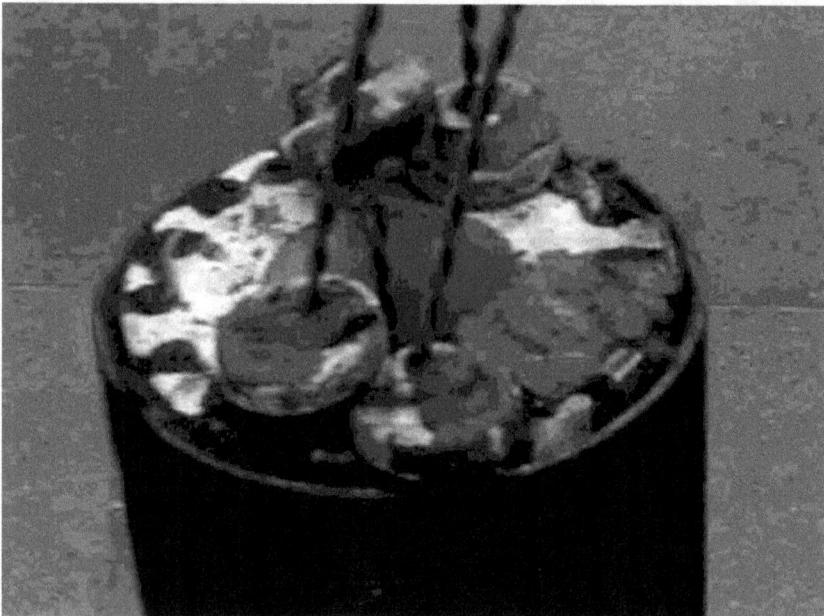

Fig. 6.4 - A particular of the reactor seen from the side of the cables.

Voltage supplied to the reactor was **gradually increased** step by step to higher values as E-Cat temperatures were shown to stabilize. A summary of the **external temperature** trend, including data collected after shutting down the E-Cat, may be seen in the chart 6.5, where we notice that the temperature is indicated in **degrees Kelvin**.

The entire test lasted 6.48 hours (from 11:21 AM to 5:50 PM). The **maximum power input** during the test has been 3.56 kW (corresponding to V = 147 V and A = 24.28 A). An average E-Cat module **temperature peak of 801° C** was measured, as calculated from Frame 17h49 (5:49:00 PM) taking room temperature into account.

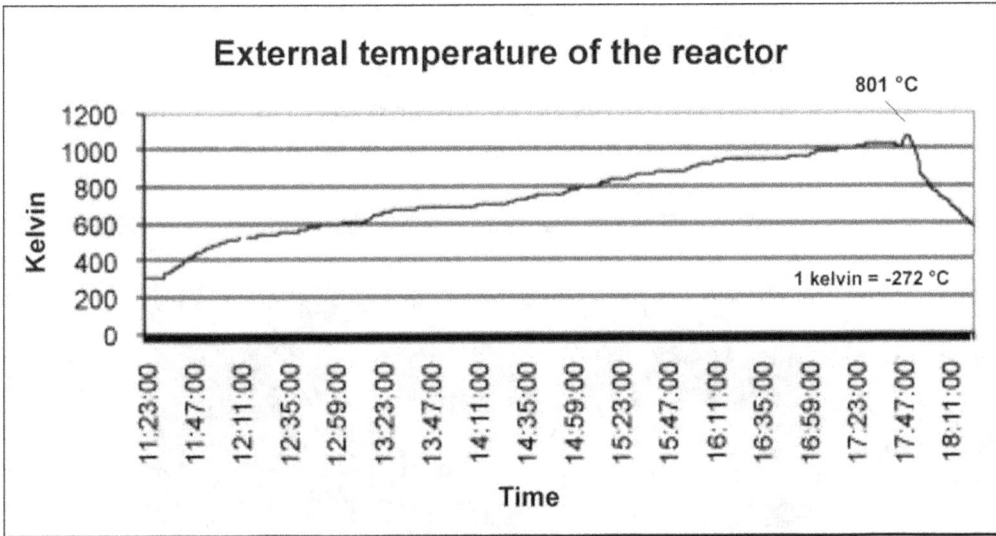

External temperature of the reactor

801 °C

1 kelvin = -272 °C

Fig. 6.5 - The external average temperature of the reactor during the test and after its shut-down.

During the test, the **internal temperature** of the cylinder – which may be seen to be glowing white hot in the photograph at the beginning of this chapter – was measured as well. In some points, the internal temperature of the cylinder **reached 1200 °C**. This measurement was performed with a laser-guided thermometer.

Some difficulties were caused by the distance, required for such a measurement, of at least 1 m from the heat source, as well as by the requirement of converging and stabilizing the beam on the entry hole. Values ranged **from 1100 °C to 1200 °C**, and were seen to increase from the entrance of the hole towards the center of the cylinder.

As pointed out by 'Cures' (it's curious that he chose not to appear in the report, being probably retired, but zooming on the image below we know that he had a major role in the test): "The **central tube** in the famous photo of this test is **empty** and white hot. If the photo had been in line, we would have seen the opposite wall! Inner wall around the white hot zone was **unapproachable** under one meter for the breath of hot air".

Penon's report focuses on the measurements performed on the E-Cat reactor for the purpose of determining the **power density** in kW/kg, and the **energy density** in kWh/kg, of the source contained within the device. Calculations of such quantities provided in the report are based on the data recorded by an *Optris PI 160* infrared camera, as well as on observations performed during the test.

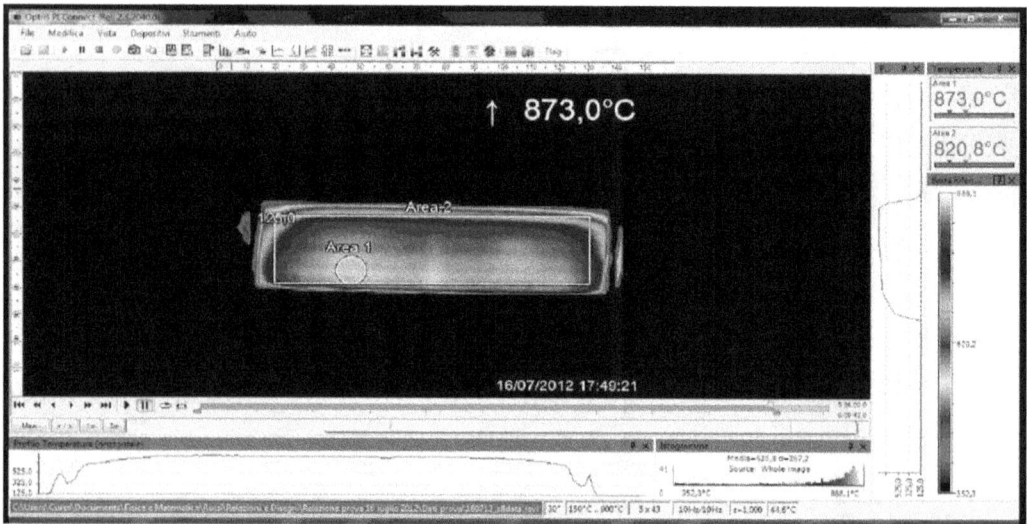

Fig. 6.6 - A frame recorded by the IR camera when the reactor was just near its maximum temperature (zooming we can read below that it was stored on a PC of the user "Cures").

Power and energy consumption during the test are summarized in the table 6.2. The **average power consumption** has been **1.28 kW**. The average power irradiated by the E-Cat module has been 2.47 kW in the worst case hypothesis that the temperature of the inner cylinder is equal **to that of the outer cylinder**, and 3.66 kW in the hypothesis that the power irradiated by the inner cylinder is equal **to that irradiated** by the outer cylinder.

Time	Difference (in minutes)	Difference (decimal fractions of hr.)	Voltage (V)	Current (A)	Power (kW)	Energy (kWh)	Total (kWh)
11:21 AM	0:00	0.00	8	1	0.01	0.001	0.001
11:31 AM	0:10	0.17	22	4	0.09	0.007	0.009
11:36 AM	0:05	0.08	33	6	0.20	0.211	0.220
12:40 PM	1:04	1.07	41	7.68	0.31	0.178	0.398
13:14 PM	0:34	0.57	55	10	0.55	0.495	0.893
14:08 PM	0:54	0.90	59	10.6	0.63	0.198	1.091
14:27 PM	0:19	0.32	65.5	11.7	0.77	0.115	1.206
14:36 PM	0:09	0.15	72	12	0.86	0.302	1.509
14:57 PM	0:21	0.35	80	14.1	1.13	0.301	1.809
15:13 PM	0:16	0.27	92	15.9	1.46	0.366	2.175
15:28 PM	0:15	0.25	100	17.2	1.72	0.831	3.007
15:57 PM	0:29	0.48	113	19.36	2.19	0.510	3.517
16:11 PM	0:14	0.23	121	20.7	2.50	2.046	5.562
17:00 PM	0:49	0.82	135	22.8	3.08	1.283	6.845
17:25 PM	0:25	0.42	147	24.25	3.56	1.485	8.330
17:50 PM	0:25	0.42					
Energy consumption (kWh)	8.330						
Total hrs.	6.48						
Average power (kW)	1.28						

Tab. 6.2 - Power and energy consumption of the reactor during Penon's test.

Thus, a **COP of 1.9 and 2.8**, respectively, can be estimated in the two hypothesis for the tested reactor. However, this calculation is **highly conservative**. Indeed, a previous analysis conducted with a dedicated software had already shown that at 1000 °C heat **loss by convection** is in the order of 8% of the total heat generated by the E-Cat cylinder and such convective loss was not taken into account in the calculations.

Tablet and charge masses were not known. The mass of the total "**active charge**" (charge + tablet) used in Penon's test was thus obtained by subtracting, from the weight of the complete module as measured at the start of the test, the weight of its inert components as **measured both before and after** the test.

Therefore, the mass of what is considered to be the active charge also includes elements that are not part of it, such as the weight of paint and sealant putty. In summary, at the end of all these measurements described in detail in Penon's report, **the mass** to be considered as **active**, or its upper limit, resulted **20.38 grams**.

Now both average power density and energy density can be calculated. In the two hypothesis mentioned above (for the average power irradiated), they result, respectively: **Average Power Density = 58 kW/kg** (1st hypothesis) or 117 kW/kg (2nd hypothesis); **Energy Density = 378 kWh/kg** (1st hypothesis) and **758 kWh/kg** (2nd hypothesis).

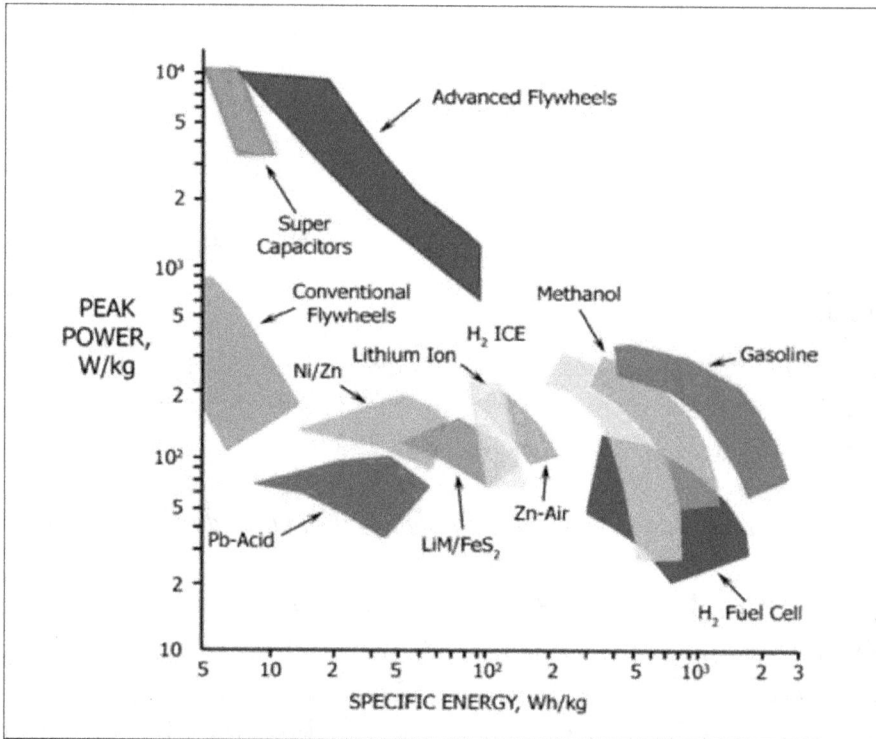

Fig. 6.7 - "Ragone plot" shows specific energy and power densities relevant to various sources. The E-Cat HT, which would be off the scale here, lies outside the region occupied by conventional sources.

In the so-called "**Ragone diagram**", energy density places the tested Hot-Cat very far from any conventional energy source. Thus, with this test, **for the first time** it has been shown that the reactions behind the invention of Rossi – the E-Cat – enters fully in the category that, only for scientific modesty, we not call "**nuclear**" but in fact they are!

CHAPTER 7

WHY THE HOT-CAT WORKS

In this chapter, I will illustrate some of what I consider "proofs" – or at least **very strong clues** – that the Hot-Cat works as claimed and is not a scam as someone suggested. I chose the ones I like the most, but I could have chosen also other examples.

The difference between a "proof" and a "clue" is based on the greater or lesser **degree of certainty**, or probability, to trace back from a known fact to one unknown: in our case, infer from experimental evidence that a Hot-Cat is really a "multiplier of energy", as declared by its "father" Rossi and by the many scientists who had it under test.

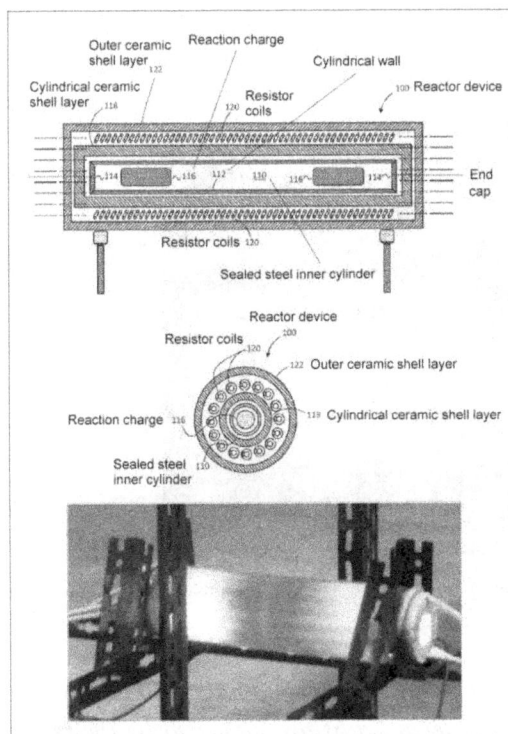

Fig. 7.1 - The E-Cat HT "Type I design" during the Third Party test performed on November 20, 2012.

In the next sections, I will illustrate the evidences based on quantitative assessments, as I am an engineer. But other people might prefer different ones. For example, those who do not like numbers, probably will consider as the most striking "proof" the figure below – published in the 1st Third Party Report (TPR-1) – showing **evidence of an exothermic reaction** that occurred within the inner tube of an E-Cat "Type I design".

Indeed, in the course of the experimental run the reactor looked just like in the picture, where you can see the shining charges placed in the two sides of the reactor and the **horizontal darker lines**, corresponding to the shadows of the resistor coils, projected outward by a source of thermal energy located further inside the device, and **of higher intensity** as compared to the energy emitted by the coils themselves.

You can see this screening also better in the IR thermal image of the same device (see Fig. 7.2), where "Area 1" is at 793 °C. The temperature dips visible in the diagram on the right of the picture are **shadows of the resistor coils,** projected on the IR thermal camera lens by a source of energy of higher intensity located inside the device.

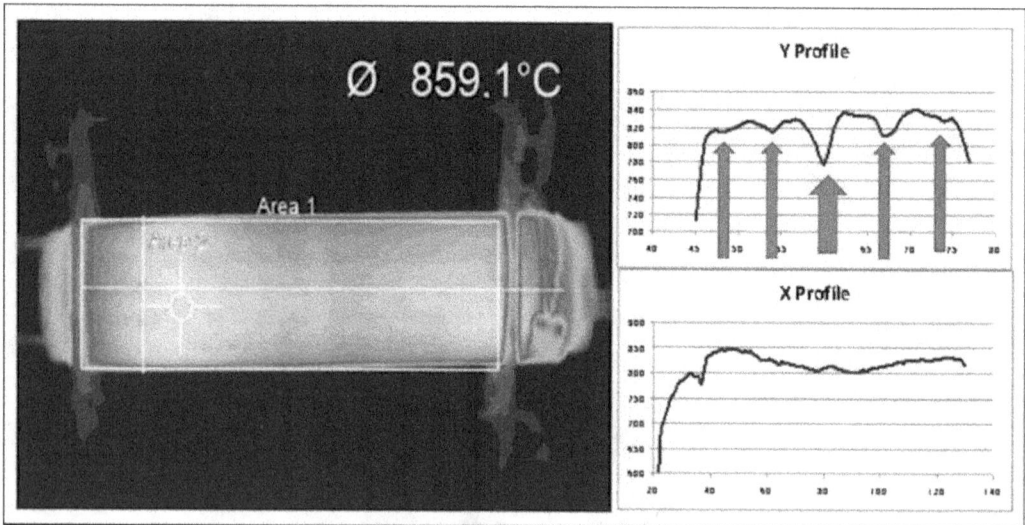

Fig. 7.2 - An IR thermal image of the November 2012 test device. Area 1 is at 793 °C. The relevant dips are indicated by the red arrows.

Please note that I decided **not to consider** too indirect evidence, such as the "replication" of a Hot-Cat by Parkhomov, resulted in a COP = 2.58. And this although he is an estimated scientist and one of the most experienced experts of LENR, and that the **"heat after death"** generated once the electrical resistor was broken – and clearly shown by its temperature graph – **leaves little doubts** that his reactor has produced a huge excess heat.

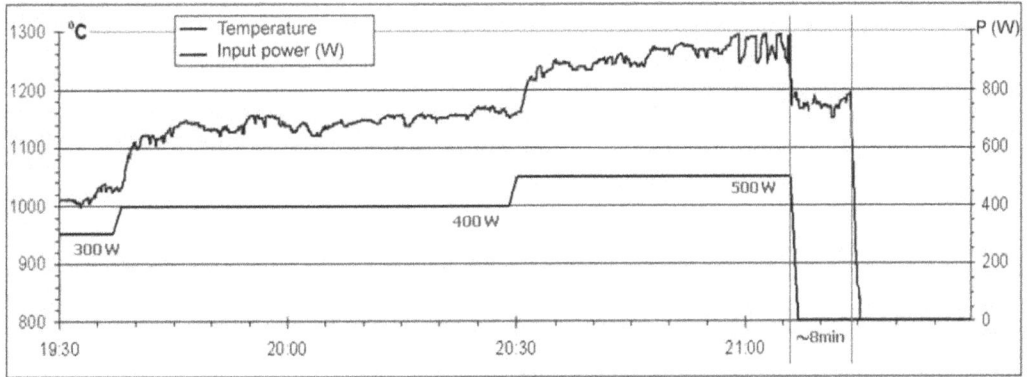

Fig. 7.3 - The 8 minutes of "heat after death" in the Parkhomov's experiment.

The big "step" in power increase

As described in the TPR-2 regarding the test of an E-Cat "Type IV design", or "Dog Bone", subsequent calculation by the authors of the report proved that **increasing** – after 10 days from the start of the experiment – the power input by roughly 100 watts (from about 800 W to 900 W) had caused an increase of **about 700 watts** in emitted power!

The picture 7.4, in which each interval on the *x*-axis represents a time span of 2 days, shows: (1) *above*: the **E-Cat Net Power Production** trend throughout the test, which is given by the difference between the total watts produced by the reactor and the watts consumed by it; (2) *below*: the **Mean Power Consumption** of the E-Cat throughout the test.

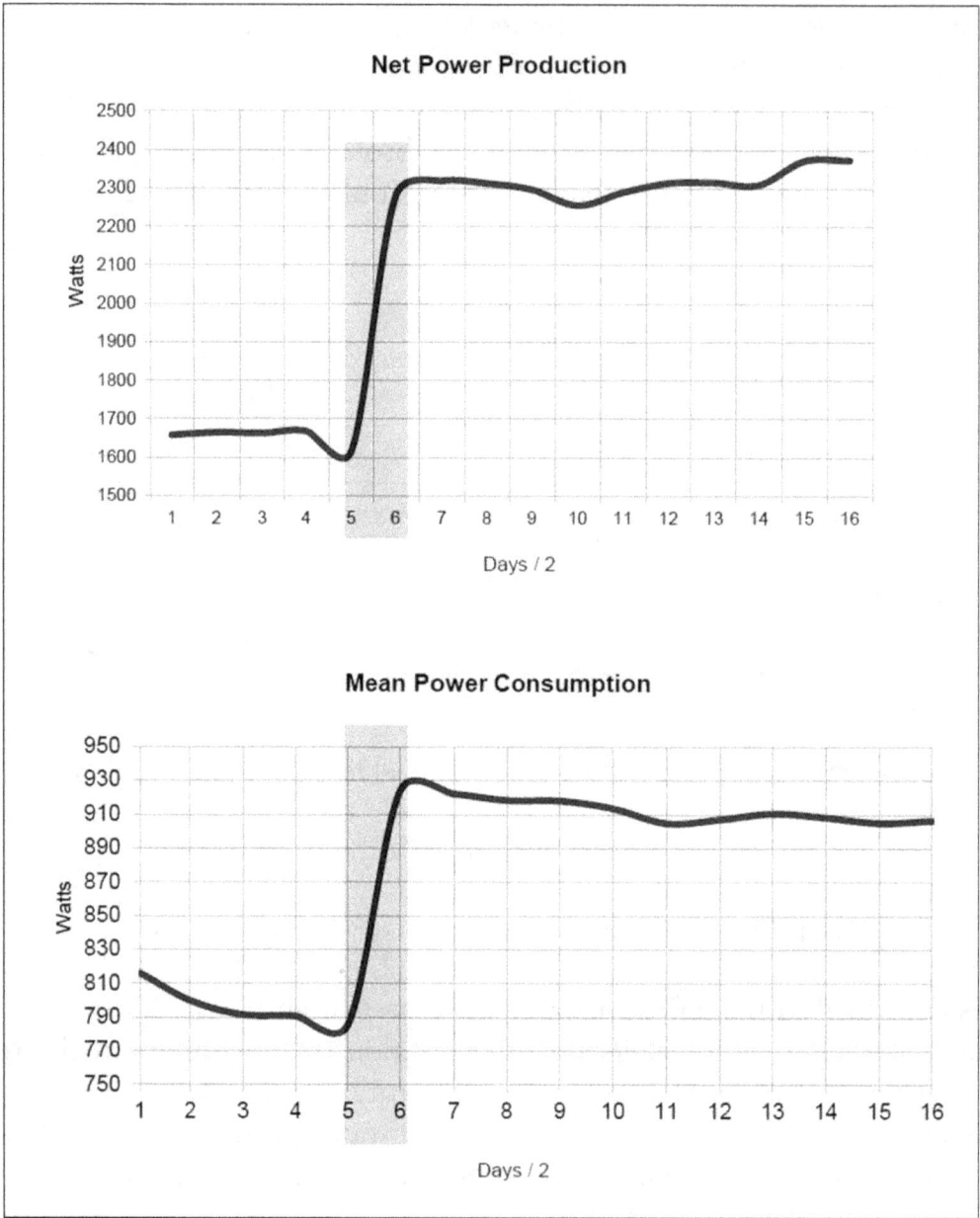

Net Power Production

Days / 2

Mean Power Consumption

Days / 2

*Fig. 7.4 - Mean Power Consumption and Net Power Production of the E-Cat
"Type IV design" tested by Third Party in 2014, as illustrated in TPR-2.*

You can see from the cited image that, when the electrical power input is increased from 780 W to 930 W (corresponding to **a step of about 140 W**), the net power production passes from 1600 W to over 2300 W, which means **an increase of** (2300-1600=) **700 W**. This huge gain is **impossible** to explain except as due to the E-Cat's internal reaction.

Indeed, if the same experiment had been performed using a **"dummy"** – i.e. a reactor with a COP = 1, or free of powder charge – then I would have expected a net power production **equal to zero** (or less than zero, due to the absorption of the control system).

So, in a dummy, a 140 W increase in the input power should produce a 0 W increase in the net output power. Even assuming (for a wrong calibration of the IR cameras) **an error** in the absolute measurements of temperature used to evaluate the output power, in any case the **corresponding step** on the chart of Net Power Production – which depends on relative, not absolute, measurements of temperature! – **could not exceed** 140 W.

A very low power consumption

During the test of the E-Cat "Type IV design" – or "Dog Bone" – described in the TPR-2, an electrical **power input** of about **900 W** has been applied to the heating resistors for a continuous period of around 20 days, producing a **power output** of about **3200 W**, corresponding to a Net Power Production of about 2300 W and to a **COP** around **3.5**.

In such conditions, the corresponding **external temperature** of the reactor resulted of about **1400 °C**, as measured through thermal IR cameras. This means that all the 2300 W of Net Power Production – and **part of the heat** that allowed the reactor to reach 1400 °C of temperature – are due exclusively to the LENR-type internal reactions.

On December 9, 2014, the *Martin Fleischmann Memorial Project* (**MFMP**) tested a reactor made of the same material, geometry and mass – and placed in a similar environment – of Rossi's "Dog Bone" (or E-Cat "Type IV design"). But it had **not been sealed well**, so the curves of temperature vs. power input obtained by MFMP during the run was, *de facto*, the calibration curves of a "dummy" reactor, with no LENR inside.

This MFMP's experiment and the relative graph (see Fig. 7.5) have clearly shown that, with an applied **power input** of about **1100 W – well above** the 900 W applied to the real "Dog Bone" in the test described in

TPR-2 – the core of the reactor is brought to approximately **1300 °C**, while the reactor's surface of alumina is brought to an estimated temperature of **900 °C** (average of the measurements of the two external thermocouples), **much lower** than the approximately 1400 °C achieved with the true Rossi's "Dog Bone".

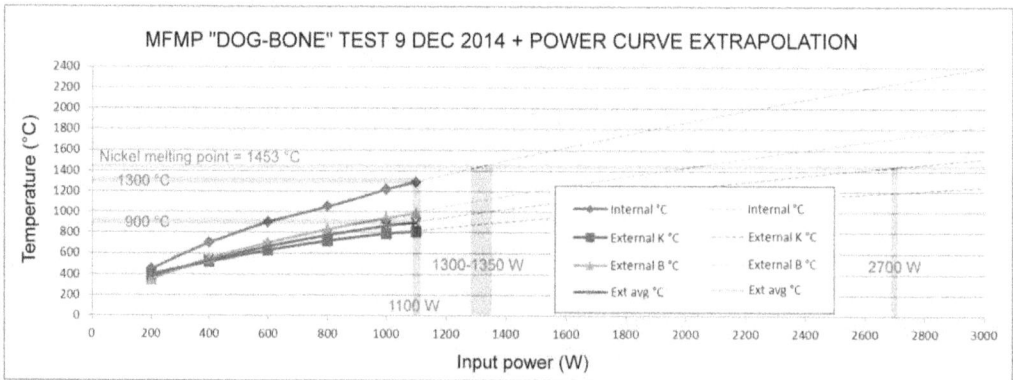

Fig. 7.5 - An extrapolation of the temperature vs. power input curves for a "Dog Bone"-like reactor tested by MFMP on December 9, 2014, but not well sealed and thus working such as a "dummy".

From the behavior of the same graph – in which the experimental curves have been suitably extrapolated for higher power inputs – it appears likely that the **core starts to melt** at an applied input power of the order of **1300-1350 W**: indeed, at these powers the temperature of the core would be higher than 1450 °C, but the **nickel melts at 1453 °C** and the nanometer nickel even at lower temperatures.

It is therefore easy to imagine what would have happened **with 3200 watts** of electrical power input applied for 32 days, that is the value required to justify a COP 3.5: if to the heating resistors had been applied 3200 W instead of the 900 W applied according to the TPR-2, nickel within the reactor **would surely melted**, because from the cited MFMP's chart we see that even the external temperature would have **exceeded 1450 °C**.

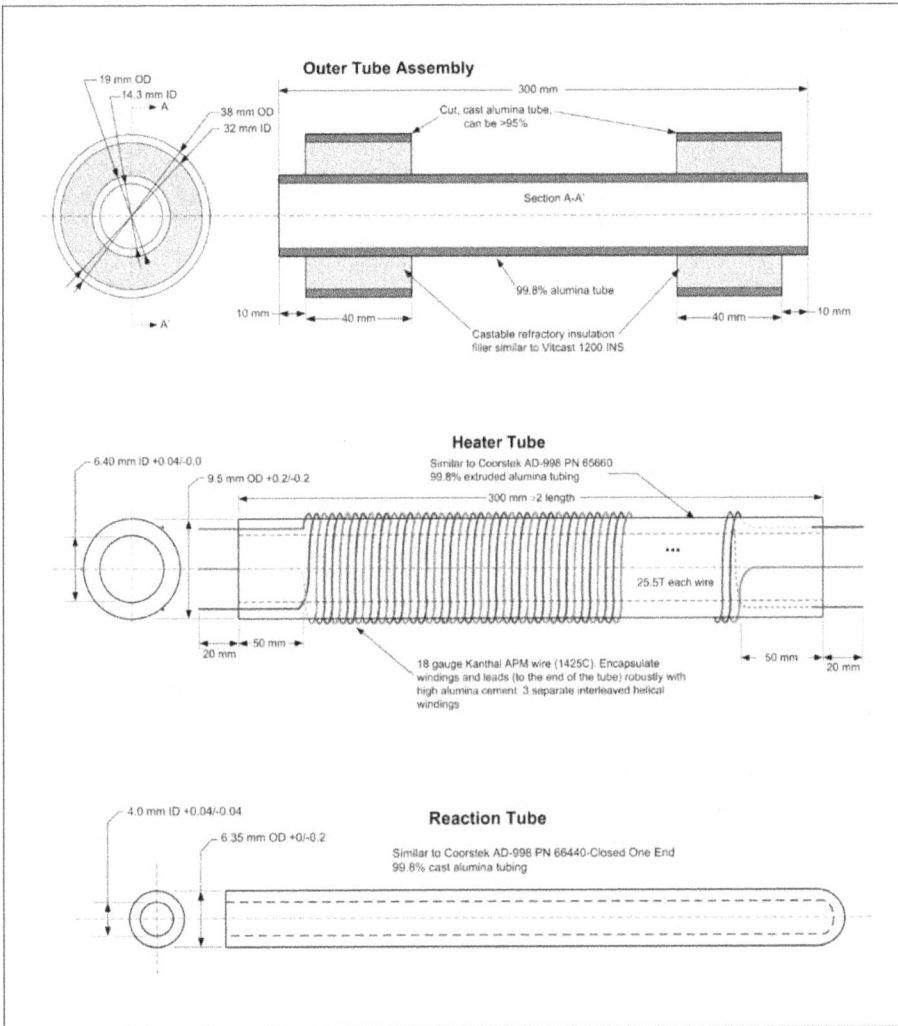

Fig. 7.6 - Assembly of the reactor designed by MFMP for their first experiment to replicate Rossi's "Dog-Bone" Hot-Cat, performed in December 2014.

The energy and power densities

It is very instructive to compare quantitatively different energy sources to see where the Hot-Cat fuel ranks. In this way, we can see that the reactions behind the operation of an E-cat are not chemical reactions, but rather **a new type of nuclear reactions**, although occurring at low temperatures and without dangerous or polluting by-products.

I found very useful, in order to make such an assessment, the chart published on the website *LENR for the Win* and given here with some slight changes. It is a so-called "**Ragone chart**", used for performance comparison of various energy-storing devices. Both axes are logarithmic, which allows comparing performance of very different devices (for example, extremely high and extremely low power).

On such a chart, the values of **energy density** (expressed in Wh/kg or, after an appropriate equivalence, in J/kg as in this case) are plotted versus **power density** (in W/kg). Conceptually, the vertical axis describes how much energy is available, whereas the horizontal axis shows how quickly that energy can be delivered, otherwise known as "power", per unit mass.

It is interesting to note that, in the *2nd Third Party Report* (TPR-2) by Levi et al., the Hot-Cat fuel showed to be **about an order of magnitude more "powerful"** than in the *1st Third Party Report* or TPR-1 (see the chart). Also renewable energy sources such as solar panels and wind plants, although imperfect candidates for inclusion in a Ragone chart, have been included for comparison.

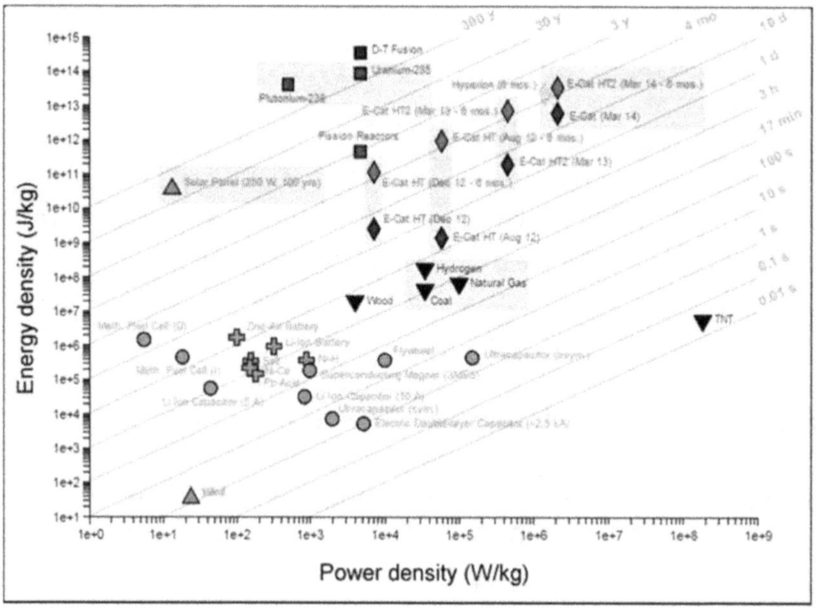

Fig. 7.7 - The Ragone Chart (from the website "LENR for the Win", slightly modified).

So, Hot-Cat fuel can be compared with a powerful nuclear source such as **Plutonium-238** and the most common **Uranium-235** (an isotope of Uranium whose fission in chain reactions is used in fission reactors), which appears to **have a much lower power density** (at least a factor 100 lower) and an energy density substantially similar.

All the traditional **fossil energy sources** (such as natural gas, coal and so on) have instead, compared with the Hot-Cat fuel, a significantly lower power density (a factor 50 lower or more) and a **much lower energy density**: at least 4 orders of magnitude lower! It is the fourth "proof" that a Hot-Cat works, and well.

CHAPTER 8

HOW THE E-CAT HT WAS INVENTED

The Hot-Cat, in reality, is a sophisticated high-temperature reactor result of a **long evolution** that began with the original E-Cat, which was a low-temperature device. It is therefore worthwhile to tell how it was invented according to my reconstruction, which will help us to understand the significant differences between the two types of reactors.

In 2011, the Swedish Defense Material Agency (FMW) has financed some very rudimentary experiments with **nickel and hydrogen**, trying to experimentally reproduce the excess heating power claimed by Andrea Rossi and Prof. Sergio Focardi (Physics Department, Bologna University) in their paper *"A new Energy Source from Nuclear Fusion"* (2010).

This information is contained in an original document 25-pages long, titled *"Experiments with Nickel and Hydrogen"*, by **Curt Edström** and **Jan Erik Nowacki**. You can find it on the web. Note that FMW, a government agency connected with the Swedish military, is the Sweden's equivalent of DARPA (Defense Advanced Projects Agency) in the US.

Fig. 8.1 - Some vessels used by the Swedish scientists in their experiments on Ni-H systems.

The report is a short description of some experiments on the nickel-hydrogen line of research, in which **4 different forms of nickel** (see the table below) were tested in contact with hydrogen at different pressures and temperatures. Some of the nickel samples also contained other metals as "catalysts" like **lithium, potassium and iron**.

In some of the samples the nickel was in micrometer large crystal grains, in other samples – for example, some **powders provided** by *Brian Ahern* upon request, because he had said to have powders that had resulted in some form of reaction – the nickel was in the form of nanometer grains embedded in **zirconium oxide**.

Type	Main metals % by weight	Grain size	From	Remark
1	Ni	100 nm	Nano Dynamics	Unknown quality
2	Ni		Israel	Carbonyl Nickel
3	Ni 90%, Fe 9%, Li<0,5%	100 nm	Own make	Noval
4a	C 4.18%, O 23%, Ni 5.4%, Si 0,21%, Zr 65%, Pd 0.86%	3 μm	Ahern EDS-analyze 1	Undamaged particle surface
4b	C 14%, O 17%, Ni 18,5%, Si 1.5% Zr 47%, Pd 1,6%	3 μm	Ahern EDS-analyze 2	Average value from many grains
4c	C 10% o 17,5 Ni 20%, Si 0.9%, Zr 46% Pd, 2,8%	3 μm	Ahern EDS-analyze 3	Very fractured piece

Tab. 8.1 - Examples of powders used in the experiments of the Swedish Defense Material Agency.

However, this document is interesting not for the description of their experiments – as **neither significant excess heat,** nor any radiation indicating nuclear reactions **has been detected** by these Swedish researchers, likely due to a not extended experimental effort – but because the group received precious **advice directly from Rossi**.

Indeed, we read already in the first page of the report: "Contact were taken with many active researchers in the field, including Andrea Rossi, for guidance to find a functioning solution. Andrea Rossi could not reveal his catalyst for us but thought that we would get **a small indicative response** using just **pure nickel and hydrogen**".

The other precious information contained in the Swedish paper is the following: "Rossi also mentioned that a hydrogen **pressure** of at least

200 bar and a **temperature** of **500 °C** was necessary in order to see any effect **without the catalyst**".

This is a news, because Rossi **had never** publicly stated that to reproduce his "Rossi Effect" the catalyst **is NOT necessary**. We have had a clue of this only from the recent Alexander Parkhomov's successful "replication" of a Hot-Cat, but the powders used by the Russian physicist include also Lithium and Aluminum under the form of **lithium aluminum hydride** (LiAlH4), so in theory one of these two elements (or both) could be a catalyst.

Fig. 8.2 - A. Parkhomov has used a LiAlH4 powder as Hydrogen source, instead of a gas cylinder.

According to prof. Sergio Focardi – that collaborated with Rossi since the end of 2007 – the function of the catalyst probably was to **transform the hydrogen** from normal, or diatomic, into **monatomic**, so that it could penetrate into the metal lattice of nickel. Therefore, it is reasonable that, at high pressures and temperatures, the catalyst is not necessary to make the hydrogen penetrate the nickel lattice, then showing a hint of **"Rossi Effect"**.

However, the words of Rossi should be taken with caution, since it is well known that at high temperatures nickel powders **sinter together**, probably making impossible the occurrence of Rossi Effect. It seems that also the Swedish researchers, in some of their experiments – probably for a deliberate choice – **did not address** this problem.

Therefore, presumably, the situation can be summarized as follows: **at high temperatures** a catalyst is not necessary, but rather an element or a substrate which avoids the sintering; whereas, **at low temperatures**, a catalyst – in the strictest sense of the term – is required to make the hydrogen monatomic.

Regarding the values of pressure and temperature suggested by Rossi to the Swedish researchers, it is interesting to notice that the first Rossi's international **patent application** for the E-Cat, filed in **2008** (WO2009125444A1), mentioned in the Abstract "a reaction between nickel and hydrogen atoms in a tube" having a pressure "preferably **from 2 to 20 bars**" and heated to "a high temperature, preferably **from 150 to 500 °C**".

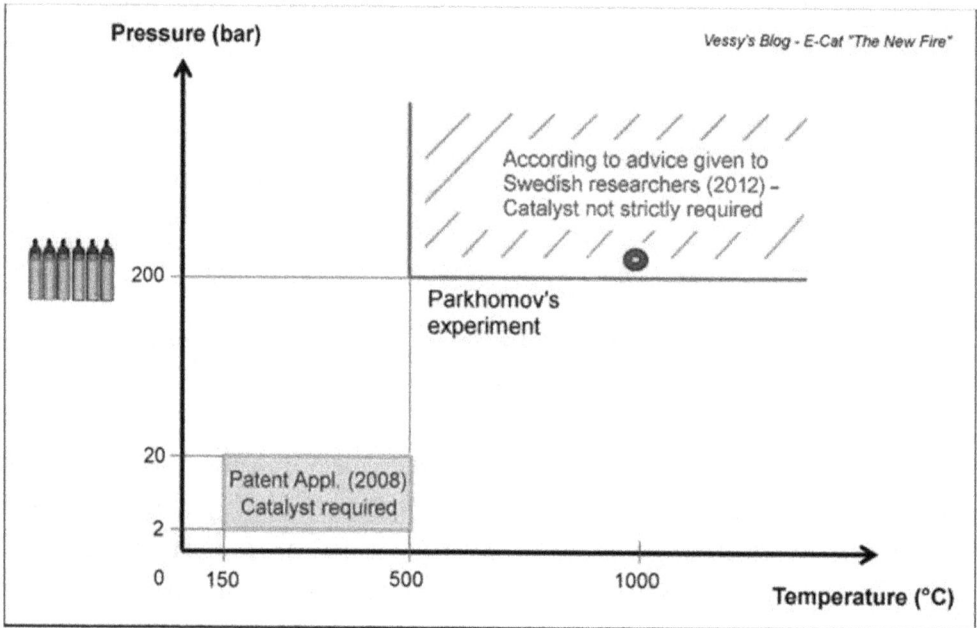

Fig. 8.3 - The two "allowed zones" for Rossi Effect on the Temperature-Pressure diagram.

The picture 8.3 shows the "**allowed zones**" for the Rossi Effect, according to the old patent application and the advice given by Rossi himself, 3-4 years later, to the Swedish scientists. I added in red the likely or reasonable position, on this Temperature-Pressure diagram, of the recent **Parkhomov's experiment**, which would explain its success.

Moreover, it's quite remarkable that **200-220 bar** is also the typical pressure of gas in Hydrogen cylinders, used by Rossi in his early experiments. So, it would seem that, in the first **MFMP's experiment** trying to replicate Rossi's "Dog Bone" reactor – but failed for an immediate degassing of the reaction chamber not well sealed – the importance of the (right) pression inside the ceramic vessel has been largely **underestimated**.

Incidentally, another important information given by Andrea Rossi is cited in the Swedish document at page 15: "As we have been told by Rossi, the powder **had to be cycled several times to get "active"** and as the power from a possible reaction would also come in 'bursts' (the length of these 'bursts' were however largely unknown to us)".

For a description of what "to be cycled" means in practice, you can read Sergio Focardi's papers on his early experiments performed (on **nickel rods** and at pressures of the order of **1 bar**) with Francesco Piantelli in Siena. One important difference: using high pressures, you do not need to use a **vacuum pump**, instead at 1 bar it was a "must".

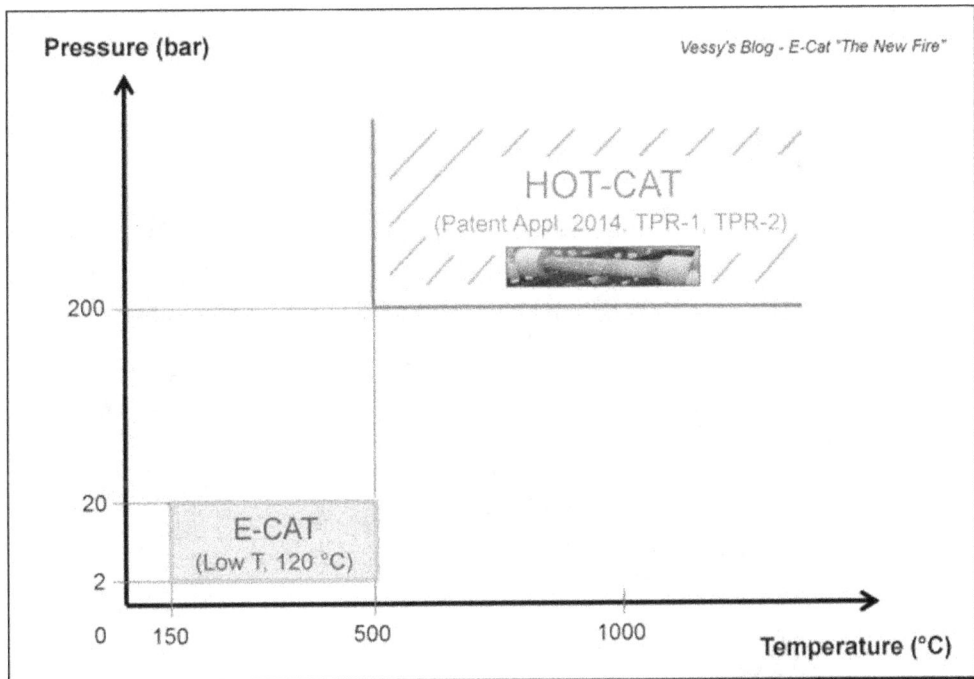

Fig. 8.4 - The two different types of E-Cat: the original, or low T, and the Hot-Cat.

From the chart we have shown in the previous pages, you can see how the two "allowed zones" for the Rossi Effect – described by Rossi himself through the first patent (2008) and his revelations to some Swedish scientists (2012) – match to the **two different types** of E-Cat (see Fig. 8.4): the original **"low temperature" version** and that, more recent, **"high temperature" (or HT)**, described in the 2nd patent application (2014) and in both TPRs.

The original, "low temperature", E-Cat was composed of a **reaction chamber** heated from **two resistors**, one internal and one external. In such chamber, once charged with nickel and a secret catalyst (likely, a couple of chemical elements), was entered hydrogen **supplied from a cylinder**. A flow of water cooled the external part of the chamber, transforming the heat into **hot water and steam**, useful for industrial or residential applications.

The most interesting feature of the original E-Cat (2008-2011) – whose fuel was already in the form of fine powders – was that it could work **self-sustaining for a very long time** the reactions with the heat generated by the reactor itself. In the 2012 version, it took approximately 30 minutes of heating with the electrical resistors to reach the correct temperature values for self-sustaining energy production.

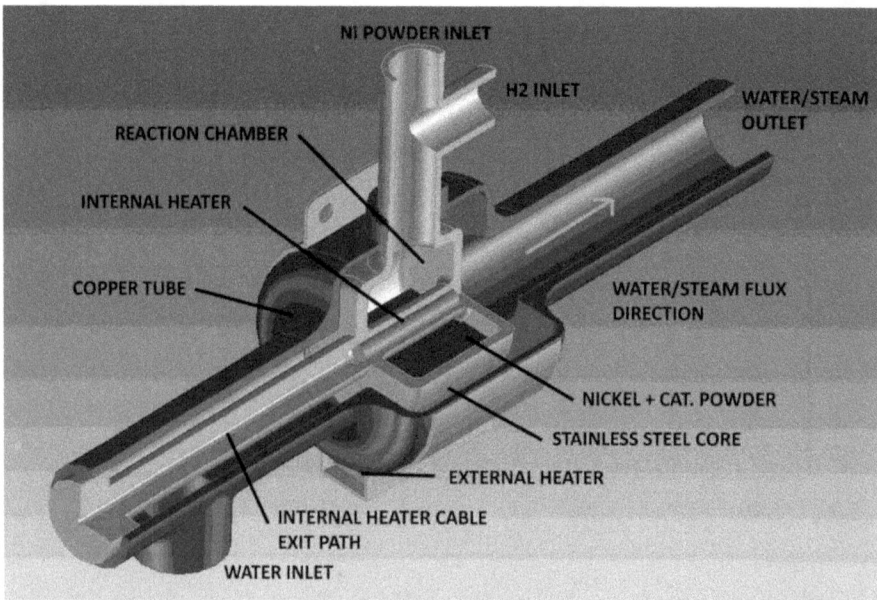

Fig. 8.5 - A speculative rendering illustrating internal core of the original "low temperature" E-Cat (courtesy Eng. Giacomo Guidi).

Also the famous **1 MW thermal plant** housed in a blue container – that you can see in many photos on the web – was simply an assembly of many small E-Cat modules of such a type **connected in parallel**. A more sophisticated version of this big, modular E-Cat could provide the required thermal energy simply turning on an appropriate number of reactors. All the operations were managed through as many programmable microprocessors.

On October 28, 2011, a **public performance test** of the 1 MW E-Cat system was performed for a Customer. Such an E-Cat could be "toggled" between self-sustaining mode requiring no power input, and powered mode running with a **COP** of **6.0**. The customer chose the first option. After a 4 hours warm up, the E-Cat worked for 5 hours **without power input**, continuously generating about **470 kW** of power output in the form of steam.

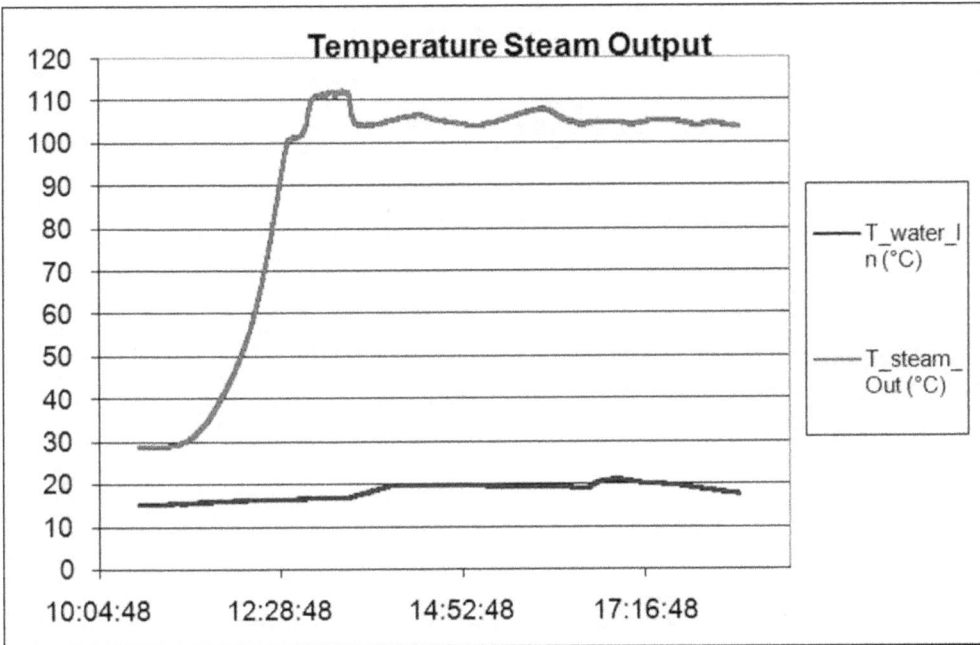

Fig. 8.6 - Chart of the output temperature during the test of 1 MW E-Cat (October 28, 2011).

Probably, after having invented the "low temperature" E-Cat – which is capable of heating water **up to 120 °C**, transforming it into steam – Rossi's team wanted to see what happened varying some fundamental physical parameters, such as pressure. Since at that time Rossi already used, as a source of hydrogen, a cylinder of gas compressed at about 200 bar, moving **from the 20 bar of the 1st patent** to 200 bar was not difficult.

Evidently, at 200 bar and at these **high heating temperatures**, Rossi – or his partner Industrial Heat, which had acquired the Intellectual Property of the E-Cat technology – found that the Rossi Effect was **faintly visible even without the catalyst** used for the "low temperature" E-Cat, but that the use of the catalyst enormously increased performance. In this way the Hot-Cat was born, but – as we shall see – it had some problems.

CHAPTER 9

SALIENT FEATURES OF A HOT-CAT

The original E-Cat and a Hot-Cat are two very different "beasts". In some sense, a Hot-Cat is a **2nd generation E-Cat**; however, it has not replaced the "low temperature" E-Cat: on the contrary, has allowed to improve it creating **two complementary products**. Indeed, a "low T" E-Cat is ideal to provide thermal energy at low temperatures (up to 120 °C) in the form of **steam**, whereas a Hot-Cat is great to provide **high temperatures**.

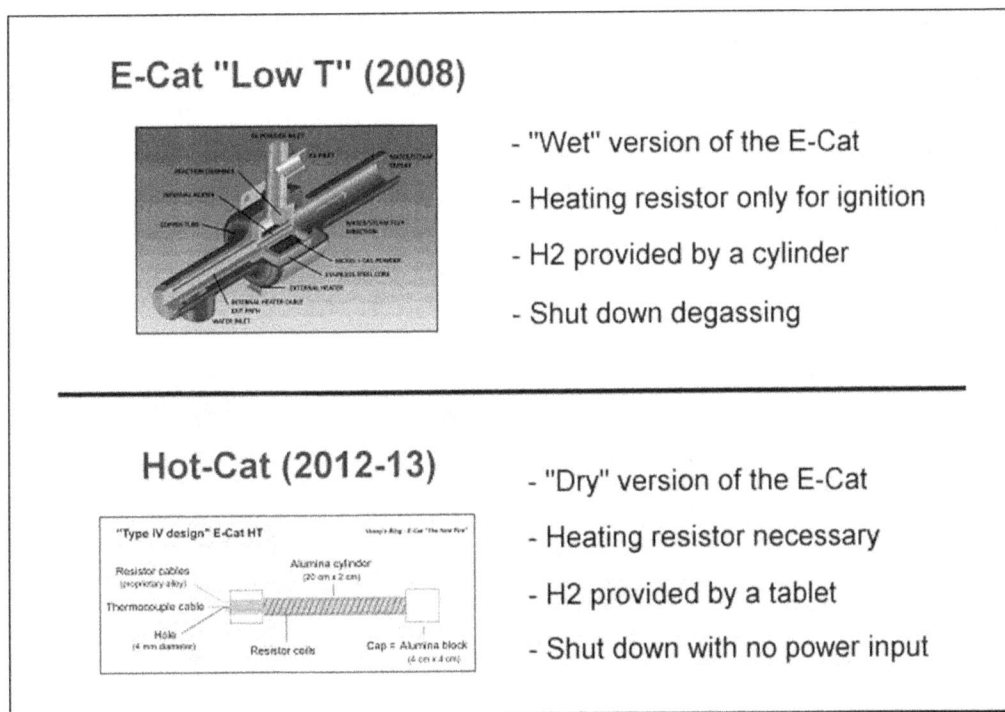

E-Cat "Low T" (2008)

- "Wet" version of the E-Cat
- Heating resistor only for ignition
- H2 provided by a cylinder
- Shut down degassing

Hot-Cat (2012-13)

- "Dry" version of the E-Cat
- Heating resistor necessary
- H2 provided by a tablet
- Shut down with no power input

Fig. 9.1 - The main differences between the original E-Cat and a Hot-Cat.

The original, low temperature E-Cat can be considered a **"wet" version** of Rossi's reactor, because it generates hot water mixed with steam. Reactions could be triggered within it by feeding an electrical heating resistor, then it was possible **permanently disconnect** the current and they continued by themselves: that is, the reactor was self-sustained until someone did not open the valve of pressurized gas to stop the process.

Instead a Hot-Cat, ideal for producing high temperatures useful for industrial processes or for generating electricity, is a **"dry" version** of E-Cat. To operate, it **needs – at least periodically – of heat** provided by one or more electrical resistors. Also the way in which the hydrogen is introduced in the reactor is different: not from the outside with a cylinder, but released by a **special tablet** once it reaches a certain temperature.

The patented tablet used by Rossi to provide this gas – probably a **metal hydride** capable of accumulating inside large amounts of hydrogen, such as Lithium Aluminum Hydride (LiAlH4), also used by Parkhomov in its successful "replication" of a Hot-Cat – also has the advantage of allowing to produce the **desired hydrogen pressure** inside the reaction chamber opportunely dosing the amount of powder in the tablet.

However, the differences between the original low temperature E-Cat described in the first patent application (2008) and its high temperature version, or Hot-Cat (2014), go **well beyond** the already mentioned. I have tried to summarize, in the table below, the **main characteristics** of both types of reactor for an immediate comparison.

Feature	E-Cat "Low T"	Hot-Cat
Year 1st prototype	2007	2012
Ignition temperature	60-70 °C	~500 °C
Output temperature	<120 °C steam	600-1000 °C
Hydrogen pressure	20 bar	>200 bar
Hydrogen source	Cylinder	Tablet
Electrical power input	Single-phase	Three-phase
No. heating resistors	2	3-many
Fuel charge quantity	A few grams	1-3 g
Type of secret catalyst	Chemical	Mix
Self-sustained mode	No power input	Cat & Mouse

Tab. 9.1 - The most important differences between an original "low temperature" E-Cat and the last generation Hot-Cat.

One of the main differences between the two types of E-Cat stands in the **ignition temperature of reactions**: for a "low temperature" E-Cat it is about 60-70 °C, as also confirmed by Sergio Focardi, whereas for the Hot-Cat it is around 500 °C, as we know from the already mentioned information given by Rossi to some Swedish scientists.

In other words, the ignition temperature of LENR reactions inside Rossi's device is **slightly lower than the operative temperature** of the same reactor, i.e. of the temperature that it is capable of providing to the water or on its external wall, respectively. Indeed, an E-Cat with ignition temperature of 60-70 °C has a working temperature up to 120 °C, and the first version of Hot-Cat had a working temperature of 600 °C.

In the two types of reactor there is also a **different power supply**. The original "low temperature" E-Cat was powered with single-phase electric power at **230 V**, whereas the Hot-Cat is always powered with a **three-phase** (the only exception was in Penon's test: two resistors coils powered at 230 V). So, the use of three-phase instead of single-phase in a Hot-Cat could not be a mere coincidence.

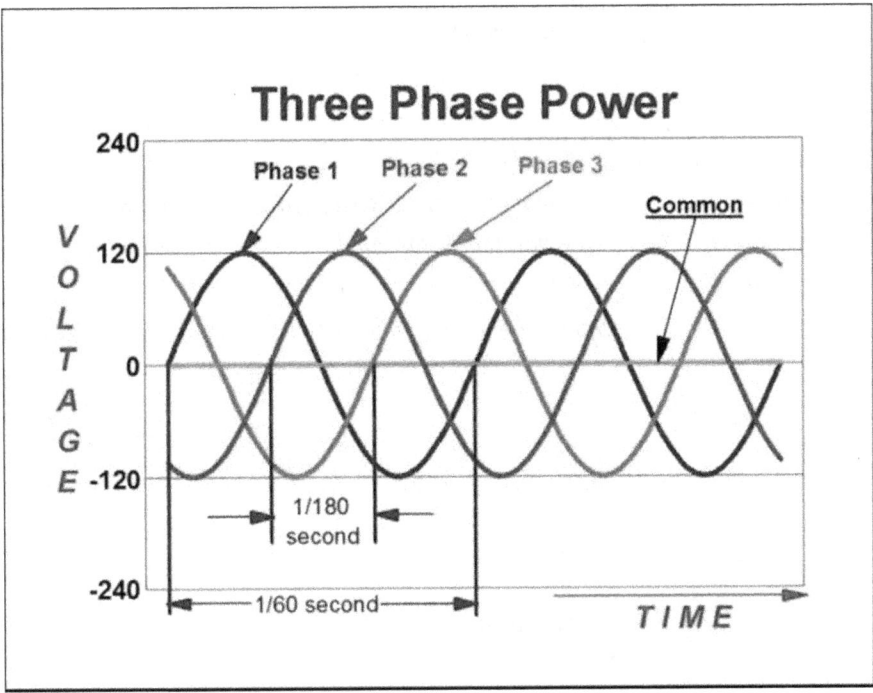

Fig. 9.2 - The time displacement of the currents in three-phase electric power.

Any **polyphase system,** by virtue of the time displacement of the currents in the phases, makes it possible to easily generate a **magnetic field** that revolves at the line frequency with a steady amplitude, provided that all three phase currents be equal in magnitude, and accurately displaced one-third of a cycle in phase (see Fig. 9.2). Also a single-phase winding produce a magnetic field, but it **cannot provide** the same effect.

In any case, the use of three-phase or single-phase electrical power instead of DC power seems to be essential. Perhaps, it could be also in some way related to the production of EMF – or **Electromotive Force** – discovered in Hot-Cats (see the following Rossi's comments posted on JoNP), that in the future may be used to **produce electricity directly,** for example by coupling the energy from a charged moving particle into a "transformer".

Andrea Rossi
December 30th, 2012 at 3:01 PM

Dear Bernie Koppenhofer:
you are touching a very important point: during these very days, and also during the more recent tests, we are working on this issue. I think we will be able to produce directly e.m.f., but much work has to be done. Actually, we already produced direct e.m.f. with the reactors at high temperature, and we measured it with the very precise measurement instrumentation introduced by the third party expert, but we are not ready for an industrial production, while we are at a high level of industrialization for the production of heat and, at this point, also of high temperature steam, which is the gate to the Carnot Cycle.

Warm Regards
A.R.

Andrea Rossi
August 7th, 2014 at 3:31 PM

The external surface of the Hot-Cat is electrically insulated, for obvious safety reasons. Currents are out of the reaction but inside the Hot Cat. If you touch *any external part of the Hot Cat you do not feel any current **nor measure any electromagnetic emission.***

Warm Regards
A.R.

Because it is **not really a force**, the term is actually a misnomer. EMF is another term for "electrical potential", or the difference in charge across a battery or voltage source. It's the potential difference in force of the electrical circuit as the current moves away from its source. Much like water flowing downhill, so is the electrical charge in an electric circuit.

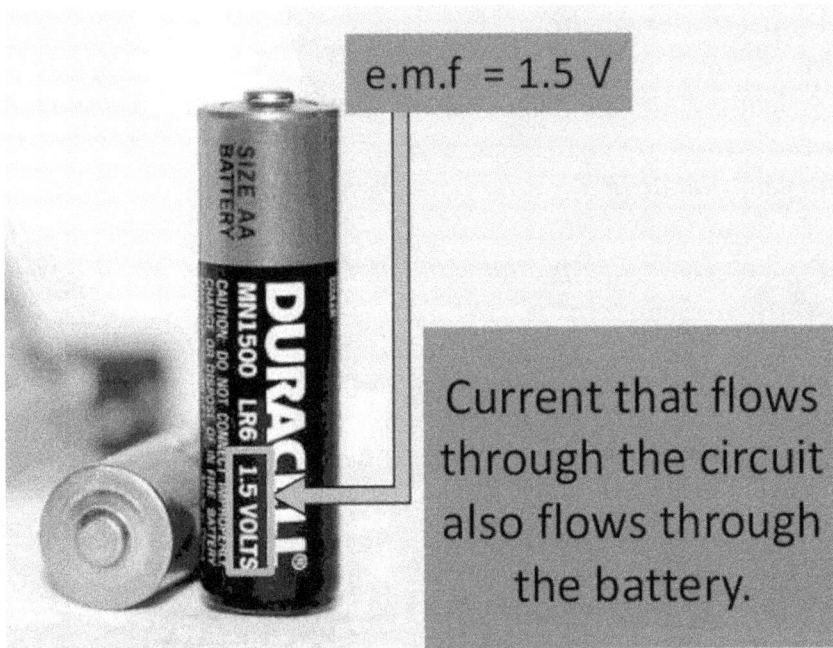

e.m.f = 1.5 V

Current that flows through the circuit also flows through the battery.

Fig. 9.3 - The Electromotive Force (EMF) exerted by a common battery.

EMF, according to Faraday's Law, represents energy per unit charge (voltage) which has been made available by the generating mechanism. So, this EMF represents a **fascinating discovery**, which gave rise to many questions. Here are some of them and the corresponding answers from Rossi posted on his JoNP in 2013-14:

Q: What form of EM fields are you measuring? Magnetic? Electrostatic?
A: electrostatic

Q: Where are they detected? Inside the reactor? Outside?
A: outside the reactor, inside the E-Cat, not outside the E-Cat

Q: Is it pulsing or constant?
A: pulsing

Q: Is the amount of EMF seen moderated or controlled by the temperatures made by the Hot Cat?

A: still under probe

Q: Will you next try to increase the effect to be able to produce electricity directly?
A: yes, possibly

Q: If this is not a temperature-dependent phenomenon, why wasn't it detected earlier?
A: matter of serendipity

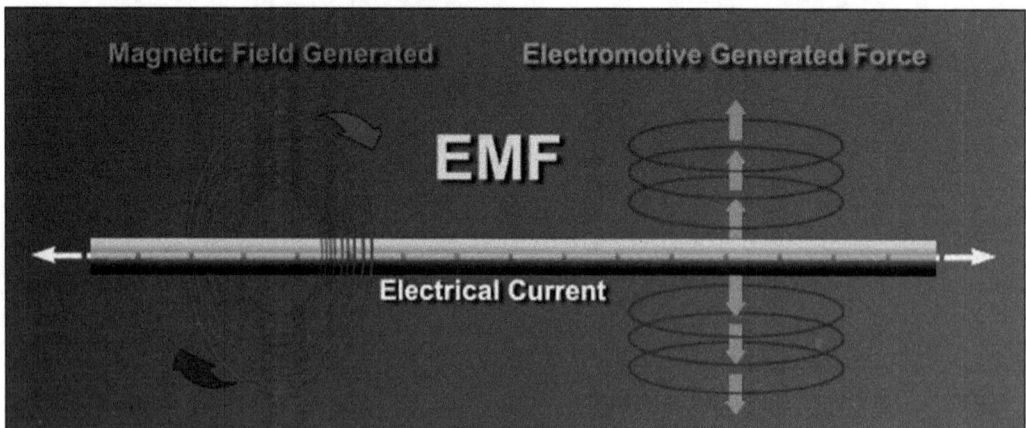

Fig. 9.4 - An electrical current flowing in a cable produces an Electromotive Force (EMF).

This "direct production of electromagnetic energy" – as Rossi call it – in the form of variable electrostatic fields is a really **unexpected** phenomenon, that makes the theoretical basis of the E-Cat even more intriguing, if they were not enough. As we can see from the picture, EMF is generated **by currents** (in our case, likely at a microscopic level), so in principle an EMF can be used to generate a current inside an electric circuit.

Towards the latest generation Hot-Cat

As "COP of the E-Cat **raises exponentially** with operation temperatures" – so that in first Hot-Cat prototypes the temperature rose too much and reactor went out of control, **exploding** – Rossi's team was forced to invent a system that would allow to drive a Hot-Cat avoiding the reactor went beyond a dangerous threshold: this system is known, in the words of Rossi, as technology of "**Cat & Mouse**".

In practice, as explained by Rossi himself in the post below extracted from his JoNP, a Hot-Cat is composed of two well separated components: an **activator** ("Mouse") and the **Energy Catalyzer** ("Cat"), i.e. the old "high temperature" E-Cat, even if probably with a different powder charge. If the Mouse excites the Cat too much, the Cat gets wild and explodes. They have seen explode hundreds of reactors in this way:

Andrea Rossi
December 27th, 2013 at 7:56 PM

Hank Mills:

*If we give too much energy to the reactor the temperature raises above the controllability limits and the reactor explodes. We must maintain the drive below this limit, and it is what we are learning to do, trying to reach a controllability level at the highest temperature possible, because **the COP raises exponentially with the operation temperature**. The apparatus is made by two well separated components, the activator ("mouse") and the energy catalyzer ("Cat"). Now we have a mouse with a COP above 1 and a Cat with a COP with zero energy consumption. If the Mouse excites the cat too much, the cat gets wild and explodes. We must not risk to reach this level. We have seen* *explode hundreds of reactors now, this way.*

Warm Regards,
A.R.

So, Rossi and his team performed many **"run away" destructive tests** in order to observe various failure modes and check for any dangerous results. This is no different than safety testing that would be done on any energy generation device. About tests to destruction, there is also another interesting comment posted by Rossi one day later on his blog:

December 28th, 2013 at 8:32 PM

James Bowery:

*Very sorry, I cannot answer to this question exhaustively, but I can say something. Obviously, the experiments are made with total respect of the safety of my team and myself. During the destructive tests we arrived to reach temperatures in the range of **2,000 Celsius degrees**, when the "mouse" excited too much the E-Cat, and it is gone out of control, in the sense that we have not been able to stop the raise of the temperature (we arrived on purpose to that level, because we wanted to study this kind of situation). A nuclear Physicist, analyzing the registration of the data, has calculated that the increase of temperature (from 1,000 Celsius to 2,000 Celsius in about **10 seconds**), considering the surface that has increased of such temperature, has implied a **power of 1 MW**, while the Mouse had a mean power of 1.3 kW. Look at the photo you have given the link of, and imagine that the cylinder was cherry red, then in 10 seconds all the cylinder became white-blue, starting from the white dot you see in the photo (after 1 second) becoming totally white-blue in the following 9 seconds, and then an explosion and the ceramic inside (which is a ceramic that melts at 2,000 Celsius) turned into a red, brilliant stone, like a ruby. When we opened the reactor, part of the AISI 310 steel was not molten, but sublimated and condensed in form of microscopic drops of steel.*

Warm Regards,
A.R.

The photo Rossi references is Fig. 9.5. After a failure, they must still "open" the reactor. So what "explodes" is not the outer container which you see in the picture, but basically the internals of the container where **ceramics and the steel have melted/exploded** due to extreme heat generation. Obviously, Rossi and his team have to make sure that such a runaway condition cannot happen on a commercial product.

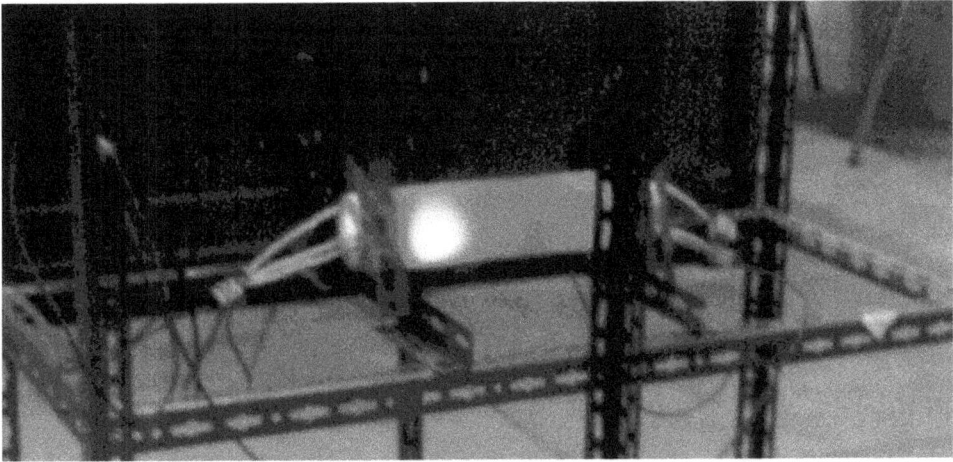

Fig. 9.5 - An Hot-Cat exhibits a hot spot during a destructive test performed in 2012.

It is also interesting the following short Q&A exchange with Andrea Rossi, about the important problem of the **hot spots** and their elimination. It was posted by Rossi on his Journal of Nuclear Physics (JoNP) on May 18, 2013:

> *Q: Have you finally found a way to supply additional heat to specific regions of the E-Cat, after these last few years of trying?* **A: Yes.**

> *Q: Are the hot spots that we recognize in the various photos of the E-Cat caused by the same phenomenon that you were trying to eliminate from the E-Cat over the last few years?* **A: Yes.**

> *Q: Do these hot spots damage the physical integrity of the E-Cat?* **A: Yes.**

If we give a careful look to the already mentioned leaked photo of a Hot-Cat **pushed to an extreme**, where there is an almost white hot region in one section of the reactor, we can understand why, according to Rossi, this is undesirable and can damage the device.

So, it is clear how the issue that Rossi has always been fighting is that of **control**. It seems like the nature of the nuclear reaction he was working with was to roar out of control until it causes physical destruction of the reactor, which of course was unacceptable. This perfectly introduces the topic of the next chapter, where we will begin to illustrate a **solution** found by Rossi for this fundamental problem.

CHAPTER 10

UNVEILING THE 'CAT & MOUSE' HOT-CAT

As already mentioned, since 2013 and for quite a long time, Andrea Rossi has described a particular type of architecture for the E-Cat that he calls **"Cat & Mouse" configuration**. He said that it was used in the E-Cat HT I've presented as "Type III design", a reactor tested by Third Party scientists in March 2013 and described in TPR-1.

Until a few months ago, there has been some confusion on the "Cat & Mouse", due to the lack of detailed and "official" explanations. Rossi provided only **vague descriptions** of such a configuration, saying that a smaller "Mouse" activates to stimulate a larger "Cat". When the Cat is stimulated, the Mouse is turned off until the Cat is no longer adequately stimulated. Then the Mouse is turned on again.

Fig. 10.1 - This image resumes quite well our knowledge of the "Cat & Mouse" configuration of a Hot-Cat about one year ago, at the beginning of 2014.

Rossi has also stated that, by using this method of periodic stimulation by the Mouse, a **start/stop or self-sustained mode (SSM)** can be utilized that allows for higher COPs. The Mouse is stated to have one type of charge and a very low COP (but greater than 1) and the Cat is claimed to have **another type of charge and a high COP**.

The following is the first comment in which Rossi mentions this new technology:

> *Andrea Rossi*
> *March 24th, 2013 at 10:53 PM*
>
> *DEAR GEORGEHANTS:*
>
> *ALL THE PROFESSORS SAID THE TEST IS GONE WELL, VERY WELL. THE LAST TEST ENDED AFTER 120 HOURS OF UNINTERRUPTED OPERATION OF THE REACTORS (**THE NEW GENERATION OF HOT CATS** IS MADE BY **A TWO STAGE SYSTEM**, MADE WITH AN **ACTIVATOR WITH RESISTANCES COUPLED WITH A KIND OF CHARGE, WHICH ACTIVATES THE E-CAT WITH A DIFFERENT CHARGE**). THE EFFECT IS STUNNING, WE SAW IN OUR PRIVATE TESTS, AND HAS BEEN REPLICATED BY THE THIRD INDIPENDENT PARTY.*
> *NOW WE PASS TO THE INDUSTRIAL ENGAGEMENTS: WE HAVE TO DELIVER OUR PLANTS BY THE END OF APRIL.*
>
> *WARM REGARDS,*
> *ANDREA ROSSI*

The differences between Activator and Reactor

Surprisingly, a much more detailed explanation of how the "Cat & Mouse" version of the Hot-Cat works has been given in the **provisional patent application** no. 61819058, filed to the US Patent and Trademark

Office (USPTO) by Andrea Rossi and his "wife" Maddalena Pascucci on 05-03-2013, but become public only in October 2014.

We can read together the complete description of the mechanism "Cat & Mouse" – and the following **analysis of the COP** for a Hot-Cat operating in self-sustained mode (SSM) using such a design – as reported in the provisional patent (pages 30-31):

"The device is provided of a first component acting as **activator** of a second component, the latter constituting the reactor proper. The purpose of the **resistor coils** is, therefore, to trigger 'Activator', which in turn – and within a certain time interval – **'ignites' the 'Reactor'**. [*We notice that in the cited provisional patent some parts appear canceled, probably because they contain important information*].

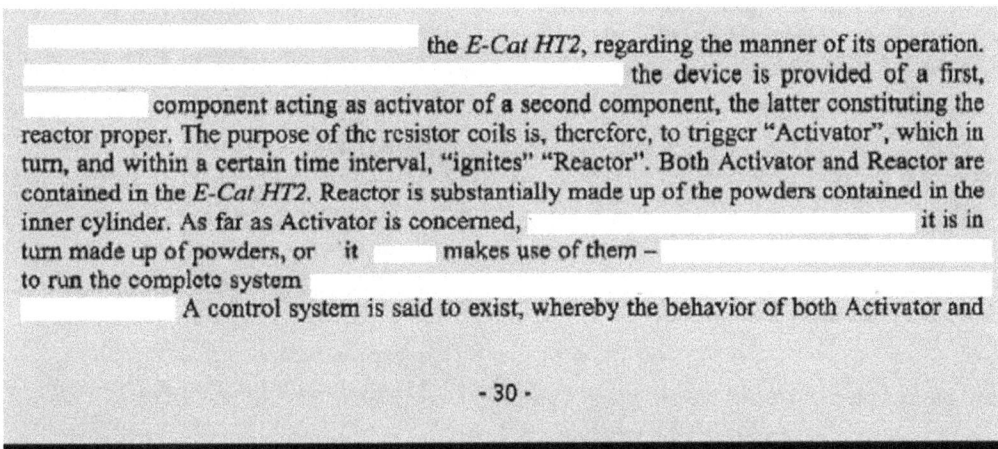

the E-Cat HT2, regarding the manner of its operation.
the device is provided of a first,
component acting as activator of a second component, the latter constituting the reactor proper. The purpose of the resistor coils is, therefore, to trigger "Activator", which in turn, and within a certain time interval, "ignites" "Reactor". Both Activator and Reactor are contained in the E-Cat HT2. Reactor is substantially made up of the powders contained in the inner cylinder. As far as Activator is concerned,   it is in turn made up of powders, or   it   makes use of them –
to run the complete system
A control system is said to exist, whereby the behavior of both Activator and

- 30 -

Fig. 10.2 - The canceled parts in the provisional patent describing "Cat & Mouse".

Both Activator and Reactor are present in the E-Cat HT 'Type III design', a model of E-Cat I've described in a previous chapter. Reactor is substantially made up of the powders contained in the inner cylinder. As far as **Activator** is concerned, it is in turn **made up of powders** – or it makes use of them – to run the complete system".

A **control system** is said to exist, whereby the behavior of both Activator and the heating curves of the E-Cat HT 'Type III design' is to be attributed to the action of Activator, the effect of which **combines** with that produced by the **resistor coils** alone, thereby accelerating the heating process".

Likewise, the temperatures recorded during the cooling phase, **higher** than those that would be got from the resistor coils alone, might be said to be 'supported' by the action of Reactor when switched off.

Even the division into three time intervals described in detail in Plot 8 may be explained in these terms. During the initial interval (I), the **resistor coils trigger Activator**: power emitted by the device is therefore lower than that absorbed by the same.

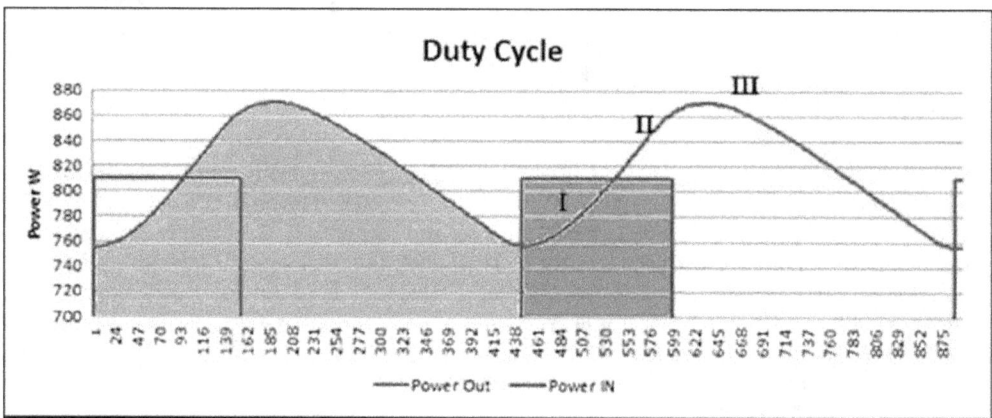

Fig. 10.3 - Chart showing emitted power (in blue) and consumed power (in red) of a E-Cat HT operating in self-sustained mode (SSM) with a "Cat & Mouse", and the 3 time intervals in which each period of a cycle may be divided.

In the second interval (II), however, the E-Cat HT 'Type III design' already emits more power than what it absorbs, indicating the **Activator itself is producing energy**.

In the third interval (III), when Activator is off – and **Reactor begins to work** – produced power reaches its peak, and subsequently falls off when the control system extinguishes the charge reaction.

If one were to evaluate these two elements in terms of instantaneous COP power, one may attribute power emitted in the first two time intervals to Activator, and power emitted in the third time to Reactor.

With reference to **one ON/OFF cycle** of Plot 8 (where emitted power values in the instants when the resistor coils are switched on and off are 759 W and 866 W, respectively, and the power input is 810 W), we get the following results:

Maximum Activator COP $_{(inst.\ power)}$ = 866/810 = 1.07

Average Activator COP $_{(inst.\ power)}$ = [(866/810) + (759/810)] /2 = 1.005

Reactor COP $_{(inst.\ power)}$ = infinite (due to the zero input energy)

If, however, one tackles the problem in terms of **energy produced and consumed**, the analysis is as follows. With reference to Plot 9, one can see that Activator COP is to be considered approximately = 1 **only** if one evaluates the first period in the ON/OFF cycle.

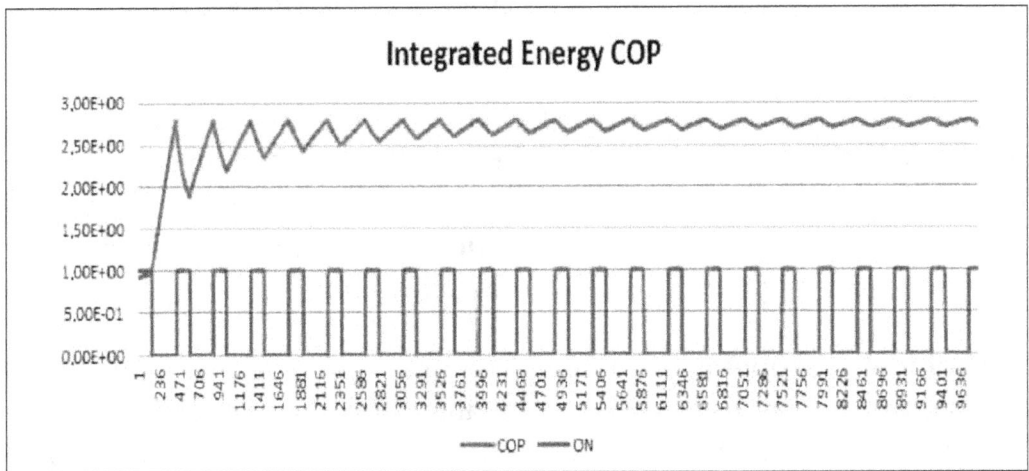

Fig. 10.4 - The blue curve is the result of the ratio between energy produced and consumed by the E-Cat HT. The red curve represents the ON/OFF trend of the resistor coils normalized to 1.

Starting from the following resistor coil ON state, the value of COP (that in this interval may be attributed only to Activator) is already approximately = 2. In the successive intervals in which Activator is in operation, one can see that its COP shows a **tendency to approach** Reactor's COP, represented by the peaks in the curve".

Here the long description provided by the provisional patent ends.

87

So, the situation illustrated in the patent for the three time intervals of a cycle should be well **resumed** by my following table, that focuses our attention on the important aspects and helps us to understand the role of the **four protagonists** of such a mechanism: (1) the resistor coils; (2) the Activator; (3) the Reactor; (4) the control system:

TIME INTERVALS	I	II	III
Resistor coils (C)	On – Triggering A	On – Triggering A	Off – Control system
Activator (A)	On – Activated by C + R	On – Much energy	Off – Controlled by C
Reactor (R)	Off – Control system	Off – Activated by A + C	On – Much energy
COP (inst. power)	Average Activator COP = 1.005		Reactor COP = infinite

Tab. 10.1 - Analysis of the actors of the "Cat & Mouse" mechanism in the three time intervals of a complete cycle.

It is quite clear, looking at this table, that the "Reactor" is activated by the "Mouse" – or "Activator" – through a **LENR mechanism**, not thermally. After all, also in the atomic pile of Enrico Fermi the runaway reactions were triggered through a **nuclear** mechanism (slowing neutrons with rods of proper materials), **not thermally**.

For **logical reasons** alone, a friend of mine suspected this for a long time before the provisional patent became public. Later, a very well informed and first-hand source **told him** that, in the last-generation Hot-Cat, there is a "Cat & Mouse" configuration and the "Cat" is not activated by the heat released from the "Mouse" but **only** by taking advantage of LENR reactions due to the Mouse.

Moreover, we have from the following comments, published by Rossi on JoNP from March to May 2013, some additional information, for example that the "Mouse" **does not use** the heat produced by the "Cat":

> *Q: Is the first stage called a 'Mouse' because it is so small?*
>
> *R: no, just to give the idea of something that wakes up a Cat: the mouse activates the aggressiveness of a Cat...*

Q: Can you feed heat back from the 2nd stage to the first stage ('Mouse') to help activate the 'MOUSE'? (…to make the mouse squeak)

R: no, are **not reversible**

Q: Does the cat and mouse both utilize separate electrical resistance based heating elements, or is the resistance element utilized by the mouse/activator the only source of external heat in the entire ECAT?

R: the only source of external energy is the resistance of the Activator

Q: When the activator/mouse is being driven or turned on, is there any power being supplied directly to the Cat?

R: no

This means that the last-generation Hot-Cat is made as shown in a picture I realized to describe the **relationships** among the protagonists.

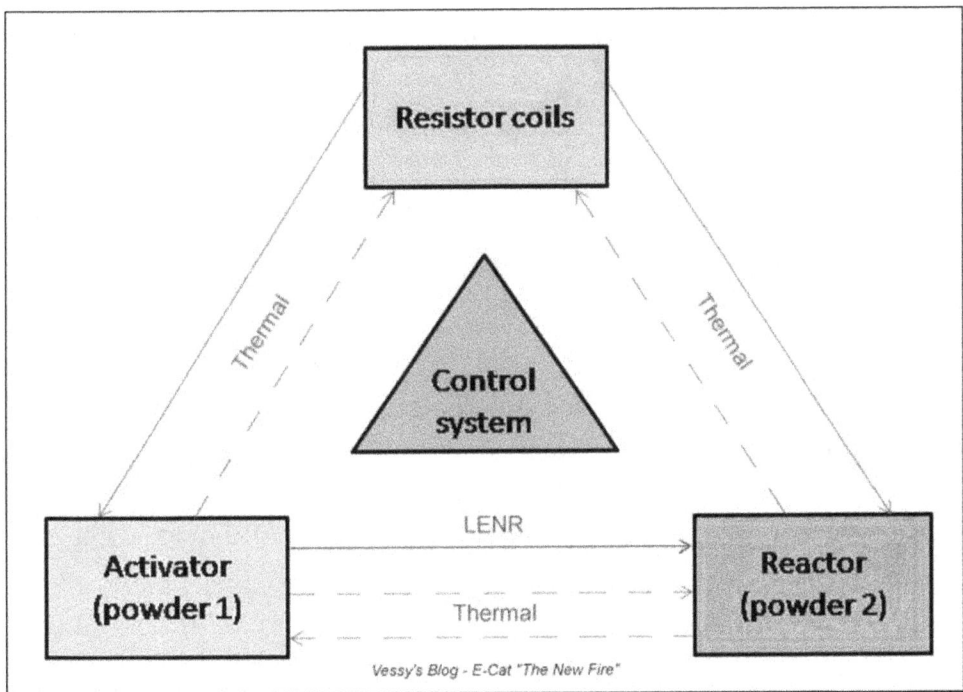

Fig. 10.5 - Likely interactions among the various components of a "Cat & Mouse" system.

In the image, we see that "Activator" is thermally triggered, whereas "Reactor" is triggered **only by LENR reactions**. The presence of the Activator allows indirectly to moderate the reactions inside the "Reactor", or "Cat", avoiding the risk of uncontrolled runaway reactions leading to the melting of nickel. In the next two chapters we'll see the advantages of this Rossi's brilliant idea applied on a last-generation Hot-Cat.

Duration of the self-sustained mode

We know that now the self-sustained mode (SSM) based on the "Cat & Mouse" technology is used both on the last-generation Hot-Cats and on the new versions of the old E-Cat "low temperature". About the **duration of the SSM**, we can read in an interview with Rossi realized by Salvo Mandarà in December 2014 (in the following text, the word "drive" is used as a synonymous of "Mouse"):

> *"The reactor can be self-sustaining for periods of time long enough. However, also **for safety reasons**, it is necessary that there is always ready to intervene an external source of energy, that we call '**drive**'. In 24 hours of operation, the system can run in self-sustained mode for about **3/4 of the time**, i.e. for 18 hours run in self-sustained mode, but for 6 hours the drive is used. When the reactor is operating in self-sustained mode, its COP is very high for obvious reasons".*

In May 2013 and, again, in December 2014, Rossi wrote on JoNP that, in self-sustained mode, the duration of the phase (in a cycle) during which the reactor does not need an electrical power input can last **up to 2 hours**, and the start-up time from a cold start was up to 4 hours. I imagine that, after such 2 hours (representing the longest duration, i.e. an upper limit) the COP reaches **dangerous thresholds**, as we'll see in next chapter.

In the five official tests on the Hot-Cat – all described in detail in the first chapters of this book – **only once** the self-sustained mode (SSM) has been used (together with the "Cat & Mouse"): in the third test described in TPR-1. In the ON/OFF phase, or SSM, the resistor coils of such Hot-Cat were powered up and down by the controller circuit at observed regular intervals of about **2 minutes for the ON** state (equal to the 35% of the time) and **4 minutes for the OFF** state (equal to the remaining 65% of the time).

Year	Month	Reactor tested	SSM	Cat & Mouse
2012	July 16	Hot-Cat for Certification (Eng. Penon)	No	No
2012	November	Hot-Cat Type I design for TPR-1	No	No
2012	December	Hot-Cat Type II design for TPR-1	No	No
2013	March	Hot-Cat Type III design for TPR-1	Yes	Yes
2014	March	Hot-Cat Type IV design for TPR-2	No	No

Tab. 10.2 - The use of the "Cat & Mouse" self-sustained mode (SSM) in the tested Hot-Cats.

So, the maximum duration of the self-sustained mode could depend on the **operation temperature** of an E-Cat: higher temperatures would imply shorter time intervals for the duration of a SSM cycle. In any case, we **must not confuse** the SSM realized with "Cat & Mouse" technology – i.e. driving the reactions through the periodic use of heating resistors – with the self-sustained mode used in the first "low temperature" E-Cats.

For example, in a 18-hours "internal" test of an E-Cat module carried out in Bologna on **February 10-11, 2011**, the reactor was heated up for 10 minutes with 1,250 W, then **it run continuously for 18 hours without power supply** (apart the 80 W required for the control unit), with a constant thermal performance of 15-20 kW allowing its self-sustaining, as witnessed by the physicist Giuseppe Levi (University of Bologna and INFN).

In a public test carried out in Bologna on **October 6, 2011**, by Eng. Mats Lewan, a small E-Cat was heated slowly increasing the electrical power input (see the chart 10.6). At 15:53 the E-Cat **reached self-sustained mode, lasted until 19:25**. The total energy produced in 3:30 hours of self-sustained mode was 10.5 kWh. The applied electric power input during SSM was 115 W, whereas those before SSM can be deduced from the chart.

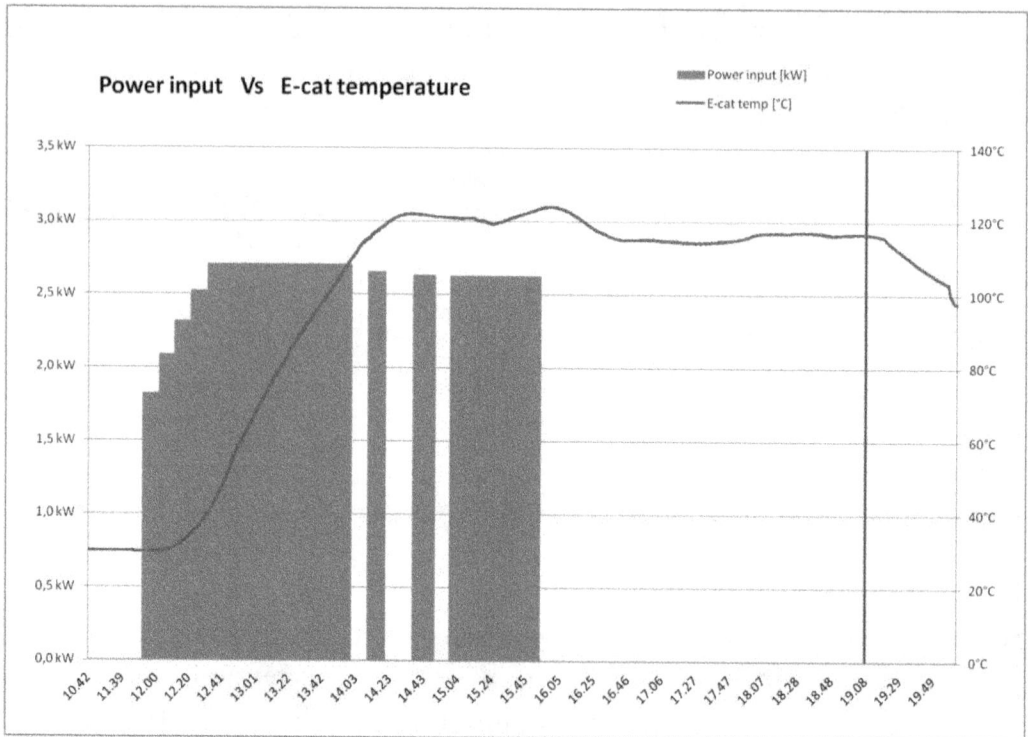

Fig. 10.6 - The test of a low temperature E-Cat carried out on October 6, 2011, by Mats Lewan. It shows the "no power input" self-sustained mode.

Also in the public test of a **1 MW** "low temperature" E-Cat performed on **October 28, 2011**, the 107 little E-Cat units 10 kW each (each module was made by 3 sub-modules of 3.3 kW each) connected in parallel and housed in a container were heated for 4 hours by electrical resistors (the reactor was turned on at 12:30), then the **power supply was eliminated** and the reactor worked self-sustaining the reactions with the heat produced (until 18:00).

In the following 5 hours after the 1 MW reactor has been turned off, the plant provided **470 kW in the form of steam**, whereas the total energy produced in the entire run was 2635 kWh, to be compared with the 66 kWh of total energy consumed, corresponding to a **COP** of 2635/66 = **40**, as certified by the tester, **Eng. Domenico Fioravanti**, who in such occasion represented a secret (perhaps military) customer.

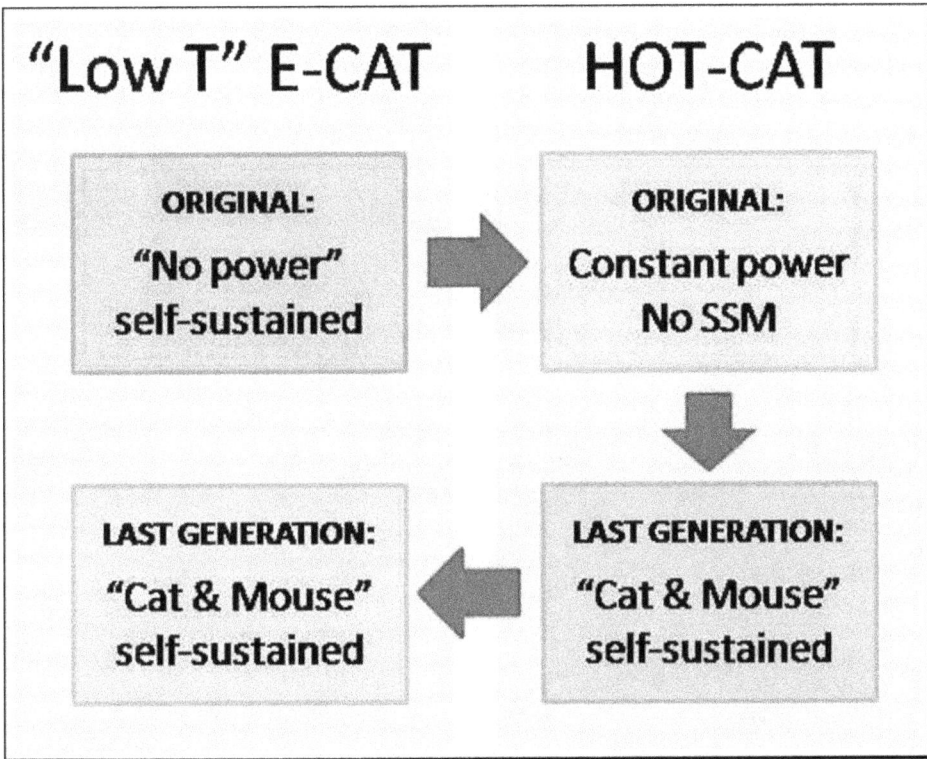

Fig. 10.7 - The evolution of both E-Cat and Hot-Cat regarding the type of power supply.

Therefore, the original, low temperature E-Cat, once ignited the reactions, could run by itself **for a very long time**: it was a **"no power input" self-sustained** E-Cat. However, more stable versions of the E-Cat – that I'll illustrate in next chapter – was later developed, such as the "zero reactivity" E-Cat and the "Cat & Mouse" E-Cat, the last being the better. Probably, the "Cat & Mouse" stability is preferred by Rossi and his team both for its **increased safety** and for eliminating the risk of "hot spots" damaging the device.

CHAPTER 11

TWO TYPES OF LONG-TERM STABILITY

As we know since the publication, in 2010, of the paper "*A new Energy Source from Nuclear Fusion*" by Rossi and Focardi, the original E-Cat "low temperature" was a device capable of an "heat output up to **hundred times** the electric energy input".

Indeed, in two tests performed in 2009 on the E-Cat, the output/input energy ratio was estimated to be around 200, thus corresponding to a **COP = 200**. But, as publicly admitted by Rossi himself, already at COP = 100 from the E-Cat there is the emission of **neutrons** (likely, slow neutrons), particles very **dangerous** for the human health. For this reason the first prototypes of E-Cat were opportunely shielded.

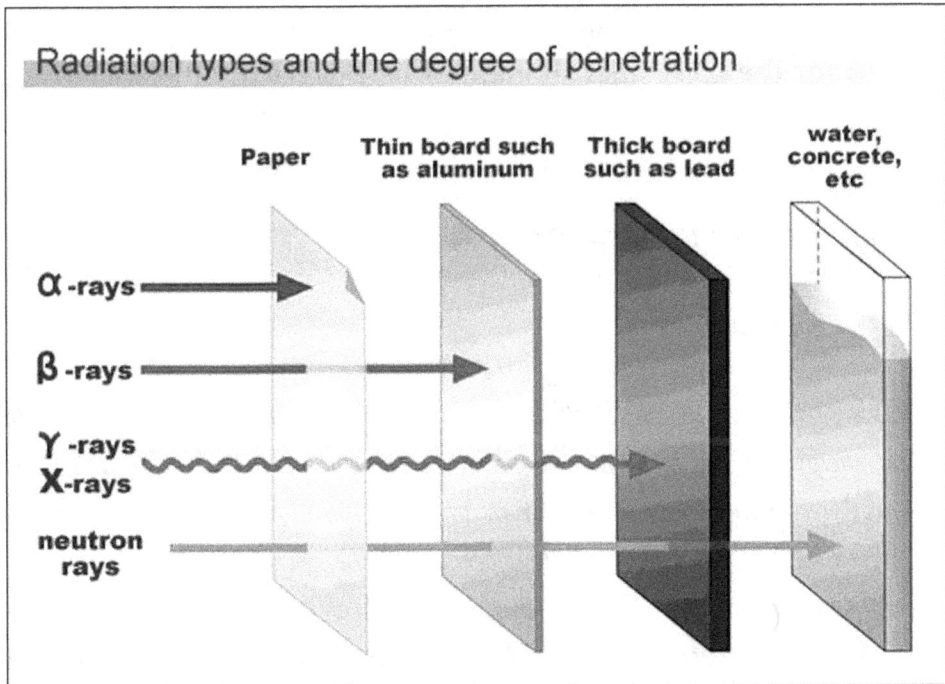

Radiation types and the degree of penetration

Paper   Thin board such as aluminum   Thick board such as lead   water, concrete, etc

α-rays

β-rays

γ-rays
X-rays

neutron rays

Fig. 11.1 - Neutrons are particles extremely penetrating and dangerous.

As Rossi said, in September 2012, during a presentation in Zurich:

"To go to a COP 200 is extremely dangerous, and when I worked at those powers it happened that we had some escape of neutrons, in some situation. For obvious reasons this is very dangerous. This is why we had to reduce to (COP) 6, and this is why we have to limit the self-sustained mode".

In this comment, Rossi refers to the duration of the self-sustained mode in the E-Cat operation. So, the danger of neutrons' escape put an **upper limit** on such parameter. This is important, because in principle the E-Cat would be able to reach extraordinarily high COPs, and only for safety reasons its COP is limited to relatively low values.

The "Start/Stop" or "Cat & Mouse" reactor stability

Also for the Hot-Cat, the achievement of a COP = 100-200 is mentioned, but **in connection** to its "Cat & Mouse" version, as you can read in the post below. So, now it seems quite clear that in both types of E-Cat – the original low temperature version and the Hot-Cat – there is a **threshold for the COP** that should not be exceeded, otherwise you risk dangerous reactions and the production of neutrons.

Andrea Rossi
May 11th, 2013 at 6:56 PM

Dear Ron Stringer:

*The Activator pays for itself, having a COP around 1.02, so its consumption of energy has not to be accounted for in the balance of the E-Cat, because it produces enough heat to pay for its own consumption with a 100% efficiency. The COP of the E-Cat is therefore net of the Activator's consumption. The COP of the E-Cat, as a matter of fact, has a zero at the denominator, but we rate it **between 100 and 200 considering some parasitic consumption of energy**, also because "infinity" makes no sense in Physics.*

Warm Regards, A.R.

All this would seem to indicate that, some time after turning off the heating resistor, both the E-Cat produce **undesirable reactions**, a side effect that must be avoided in any way. Probably the "Mouse" performs this protective function, particularly important in the Hot-Cat, because we know that COP increases **exponentially** with temperature.

In other words, the periodic electrical supply of the heating coils and the resulting activation of the "Mouse" would produce a **control effect**, preventing the nuclear reactions from running away. But how this control effect is performed is still a matter of speculation, being likely the most important **secret** of this device, together with the catalyst.

The most obvious explanation would be that the resistor coils and the "Mouse" not only stimulate the nuclear reactions in the "Cat", but also **reset the previous situation** in the "Cat" every time they are turned on. In this way, it is as if the reactions in the "Cat" started again from scratch – maybe **in different areas and/or with different intensities** – preventing the triggering of both runaway processes and dangerous reaction products.

This simple and "natural" explanation would imply that if – for a deliberate choice or for accident – the resistor coils and the mouse are not turned on, after a certain time the "Cat" **runs out of control**. But there are two clues that indicate otherwise.

The first is that Rossi, in response to a specific question on this issue made some time ago by a friend of mine, replied that, if the mouse is no longer turned on, the "Cat" **does not go out of control**. The second clue is that, otherwise, it would be impossible to shut down the reactor, because – as far as we know – **there is not another direct mechanism** to do that.

In the original E-Cat "low temperature", for example, the reactions were stopped by simply **opening a valve** to vent away the hydrogen, and thus lower the pressure in the reaction chamber. But in the Hot-cat – and also in the new "low T" E-Cat – the hydrogen is released from a patented tablet containing a chemical compound that **reabsorbs hydrogen** when the temperature in the reactor is lowered. Therefore, there is no adjustment valve.

On the other hand, when in a "Cat & Mouse" Hot-Cat the resistor coils are turned off, the temperature **starts dropping** until the resistance coils are turned back on, even it is not falling as fast as it would do in a dummy reactor, due to heat being produced by LENR. So, it seems a Hot-Cat stays **stable simply by letting the reactions die off**, instead of trying to keep them stable and continuous, as was made in previous Hot-Cat versions:

> *Andrea Rossi*
> *March 25th, 2013 at 8:03 AM*
> *Dear Steven N. Karels,*
>
> *The new technology of the Hot Cats* **has revolutionized the system we had used before.** *Please wait the publication, afterward we will talk of this issue. Basically, we have no more a reactor or a cluster of reactors, but each reactor is* **coupled with an activator:** *let's make this model: the activator is the Mouse, which makes the Cat run. The Mouse has his own COP which is more than 1, the Cat is a surprise.*
>
> *Warm Regards, A.R.*

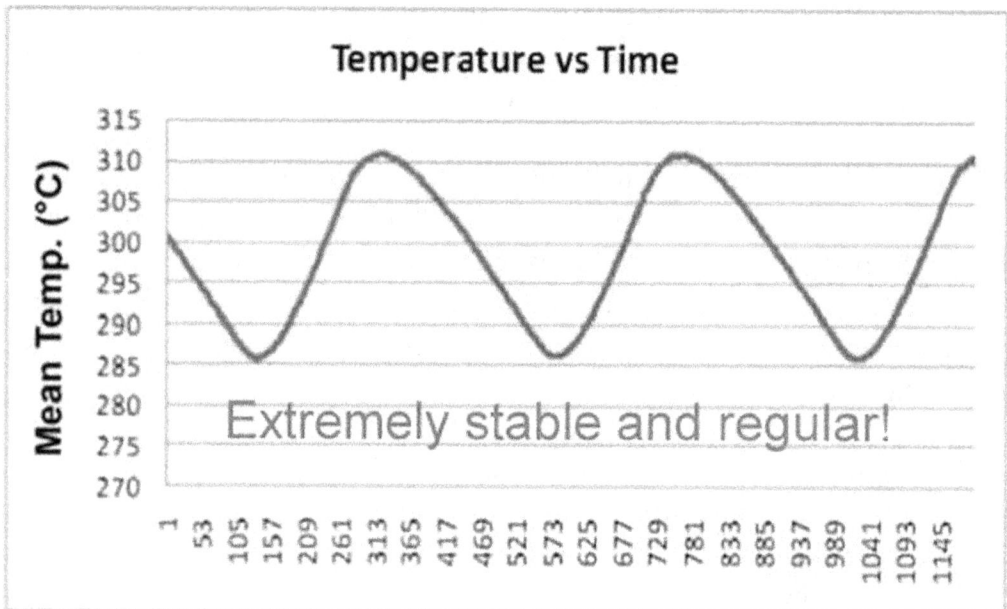

Temperature vs Time

Mean Temp. (°C)

Extremely stable and regular!

In Fig. 11.2, we can see very well the **stable and regular oscillations** of the average surface temperature of the Hot-Cat "Type III design" working in "**self-sustained-mode**" (SSM) with the "Mouse & Cat" technology. One can clearly see how the temperature of the reactor varies between a maximum and a minimum value with a fixed periodicity.

It is interesting to note the heating and cooling trend of the device, which appears to be **different from the exponential trend** typical of generic resistors. Indeed, if we compare in detail the curves of the heating and cooling phases of the device with the **standard curves** of a generic resistor, we observe that the former differ from the latter in that they are not exponential, as shown in Fig. 11.3 taken from TPR-1.

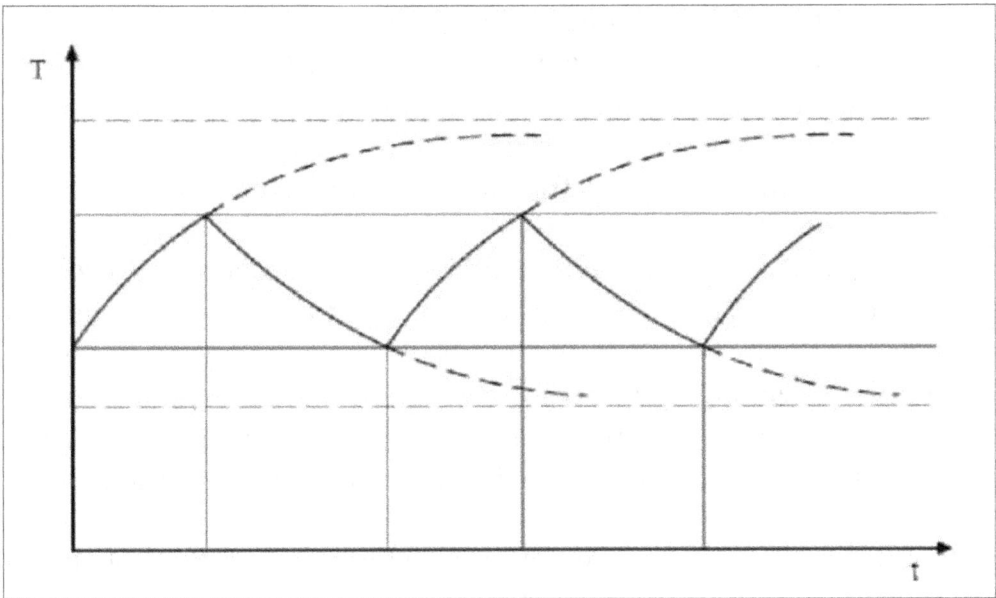

Fig. 11.3 - Heating and cooling cycle of a generic resistor. The trend is described by exponential type equations.

The alternative: a "zero reactivity" reactor stability

From the Lugano report, we know that the tested reactor was shut down **acting on the heating element**, i.e. the resistor coils: "About 32 days from startup, on the 29 March 2014, at 11:40, the E-Cat was shut down, after gradually reducing its power input". In other words, the "**deactivation**" of a Hot-Cat is the opposite of its "activation" (made at the beginning of a run), which requires a **gradual increase** in the input power.

But in the Hot-Cat tested in Lugano no self-sustained mode (SSM) was used, and later Rossi said that only **one kind of powder** was used and that "when start/stop or self sustain mode is not used, the difference between the Mouse and the Cat vanishes". However, keep a reactor **at a constant temperature** for days and days is not simpler than using a "Cat & Mouse" strategy, because a **perfect balance** of many factors is required.

Indeed, from nuclear engineering we know that "**reactivity**" is the name of the quantity which determines the rate of change of reactor power. For a reactor power to be steady the reactivity must be zero. Maintaining zero reactivity requires a good knowledge of all the parameters involved in the process, including materials, reactions, products. This allows to solve the problem of a **LENR-thermal stability** in a reactor device.

The reactivity of a Hot-Cat may be adjusted by the **reactor control system** in order to obtain a desired power level change or to keep the same power level. It can be compared to the reaction of an automobile as conditions around it change (for instance, wind intensity and direction or road slope), and therefore the corresponding counter-measure that the driver applies to maintain road speed or execute a desired manoeuvre.

In the Lugano report there are not charts showing the **trend of the external temperature** of the reactor during the entire test, but only those illustrating the net power production and the mean power consumption of the tested Hot-Cat. However, the **analysis of the results** regarding the trend of net production vs. that of consumption is truly interesting.

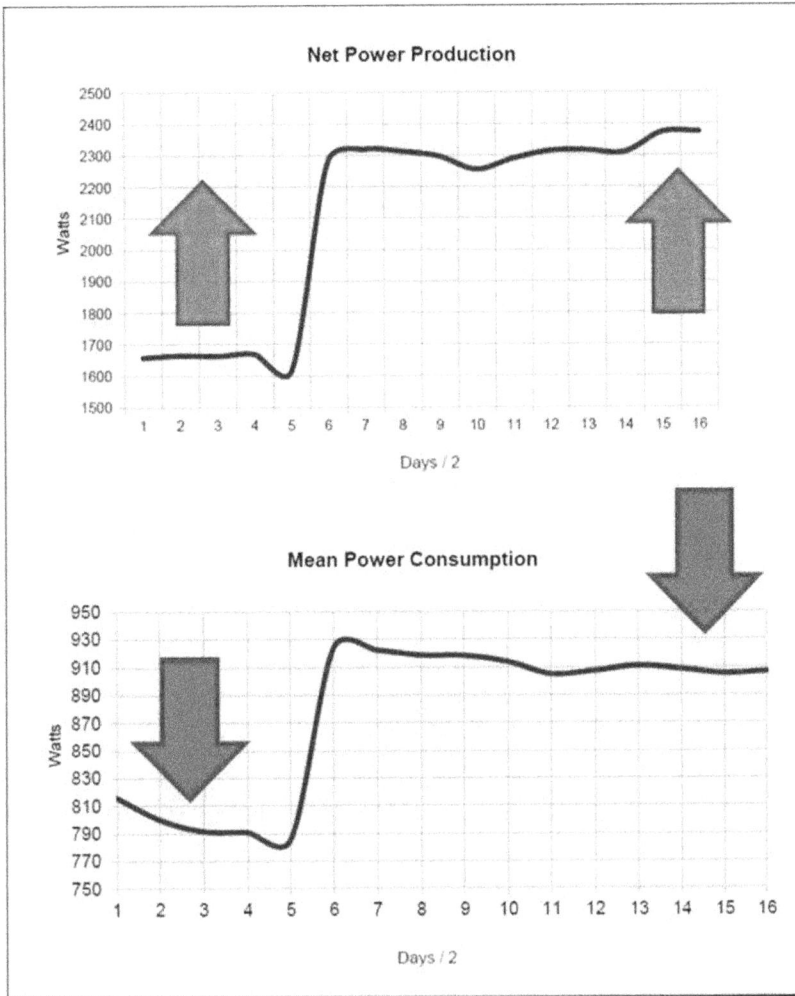

Fig. 11.4 - The net power production and the mean power consumption of the Hot-Cat "Dog Bone" throughout the entire Lugano test.

According to the authors of the report, there seems to be an **anticorrelation** between the two behaviors, which stands out as a decrease in average consumption values corresponding to increases in production averages, and vice versa. This behavior would probably due to a **feedback effect** driving the resistor power supply, raising it or lowering it according to the internal temperatures read by the thermocouple.

I'm **not sure** that their conclusions are correct because in the central part of the chart such an anticorrelation is not present. However it's reasonable that this active drive through the check of the parameter "temperature" – and consequent adjustment of the power input level – finally results in the **"zero reactivity" stability**.

So, both **temperature** and **electrical power input** would strongly affect the control of reactivity on this type of reactor, as apparently shown from the E-Cat behavior in some parts (indicated with red arrows) of the mentioned chart. Of course, these parameters seem to be **linked**, because raise or lower the power input leads, respectively, an increase or a decrease of the temperature in the reactor.

Although it's somewhat incredible at first, a Hot-Cat can be stable and yet to be **at any desired power level**. We have seen this very well in occasion of the Lugano test, when after 10 days the power input has been raised from an almost-constant level to another almost-constant level, where the "almost" refers to the small adjustments feedback driven.

In the Lugano report, there is a chart showing the fluctuations in the external temperature of the Hot-Cat (in one central area of the 10 monitored by the IR camera) during the short transient phase in which the power input was increased. We can see **how long it took** the E-Cat to stabilize at the new power level after input current was increased: this amounts to about **400 seconds**, slightly more than six minutes.

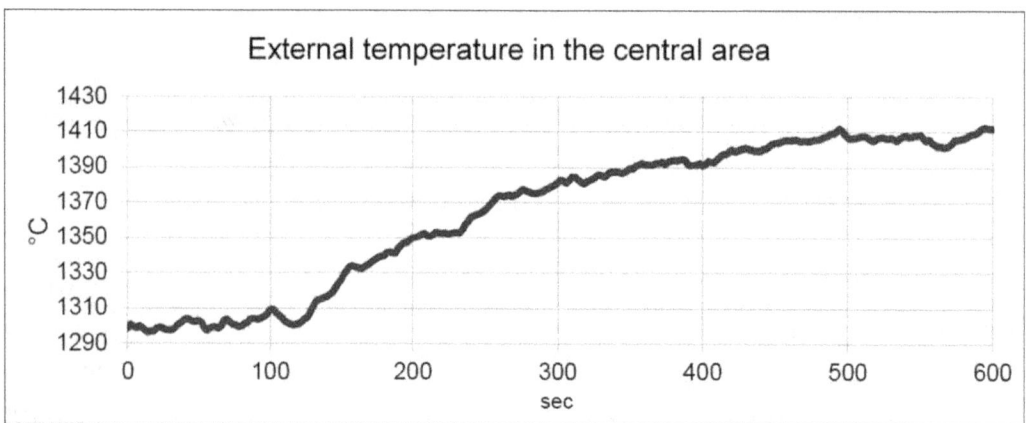

Fig. 11.5 - Average temperature in a central area of the Hot-Cat at the time of power supply increase in the Lugano test, around 10 day from its beginning.

All these interesting information give us an idea of the hard work made on the Hot-Cat. Already in 2013, Rossi wrote on his JoNP the comment that they were performing extended tests of a Hot-Cat running at 1000 °C. The improvements they were testing included trying **to raise the stable operational point** of the Hot-Cat as high as possible. Rossi indicated that they were running stably at this temperature.

But **we don't know** if, at that time, they were working mainly with "zero reactivity" reactors or with the new "Cat & Mouse" self-sustained mode. However, the "zero reactivity" stability (indicated with "0-R") was used in the major part of the tests performed in the period 2012-2013, having the self-sustained mode (SSM) been used **only** in the test performed in March 2013 (see Tab. 11.6), i.e. in the third test described in TPR-1.

Year	Month	Reactor tested	Mode	COP
2012	July 16	Hot-Cat for Certification (Eng. Penon)	0-R	1.9-2.8
2012	November	Hot-Cat Type I design for TPR-1	0-R	---
2012	December	Hot-Cat Type II design for TPR-1	0-R	5.6
2013	March	Hot-Cat Type III design for TPR-1	SSM	2.9
2014	March	Hot-Cat Type IV design for TPR-2	0-R	3.2-3.6

Fig. 11.6 - The operation mode used in the various official tests performed on Hot-Cats.

As an example of the trend of the thermal power emitted by a Hot-Cat in a situation of "zero reactivity" stability, we can see a chart (Fig. 11.7) showing the **radiative thermal power during the entire test** of a Hot-Cat "Type II design", performed in December 2012 and described as second test in TPR-1. I would point out the **good stability already achieved**, even if there were still some drifts and fluctuations, especially in the final part of the run.

In the mentioned chart of the radiative thermal power of the third Hot-Cat tested for TPR-1, it is interesting to note the **similar behavior** in the two points indicated by the blue arrows: the first after about 10 hours of operation and the second after about 83 hours of continuous operation. However, it's not clear, from this a few data, if this is a characteristic and recurrent pattern for this **"zero reactivity" operation mode**.

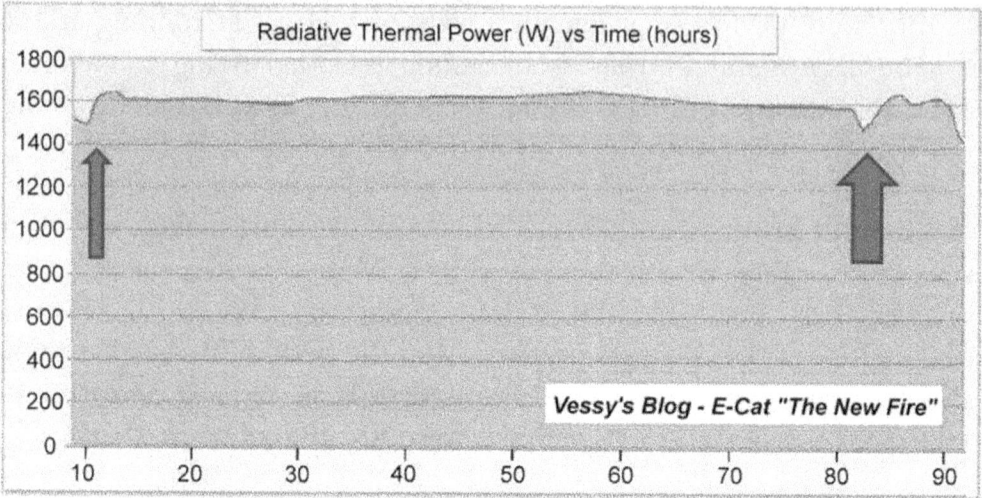

Fig. 11.7 - The radiative thermal power emitted by a Hot-Cat working in a "zero reactivity" mode.

CHAPTER 12

CONTROL OF HOT-CAT REACTIONS

The long previous discussion allows us to advance now a reasonable explanation on how the Hot-Cat based on the "Cat & Mouse" technology works. Such a Hot-Cat could be a "**coupled-core reactor**" consisting of two independently "**subcritical**" cores. The LENR coupling between them would make the entire system "**supercritical**".

When in a traditional **nuclear fission reactor** the neutron population – which allows the fission reactions that produce excess energy and also radioactivity – remains **steady** from one generation to the next (creating as many new neutrons as are lost), the fission chain reaction is self-sustaining and the reactor's condition is referred to as "**critical**".

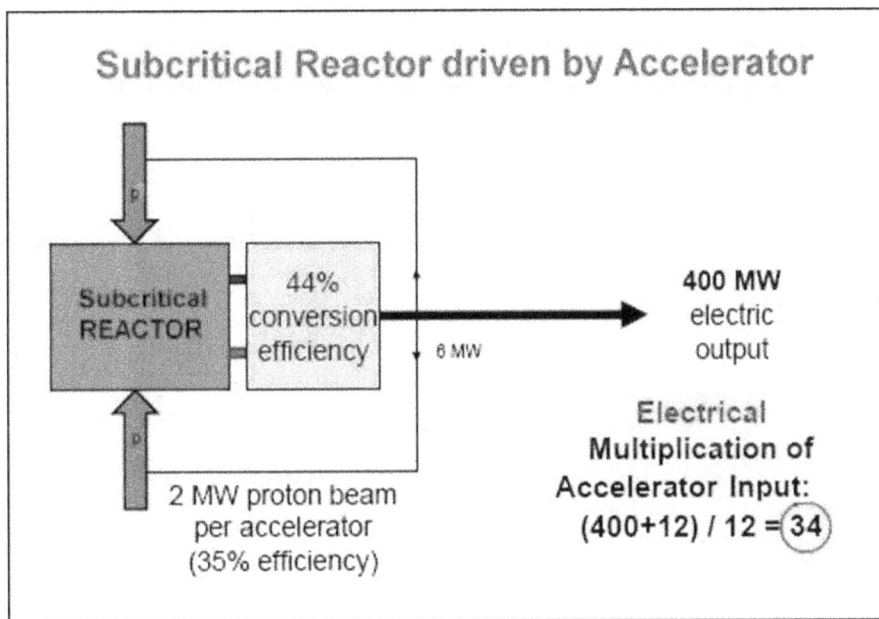

Fig. 12.1 - GEM STAR, an example of super-safe subcritical nuclear reactor driven by an accelerator, which pumps neutrons into a subcritical fissionable fuel.

But when, on the contrary, the reactor's neutron production **exceeds losses**, characterized by increasing power level, it is considered "supercritical"; and, finally, when **losses dominate**, it is considered "subcritical" and exhibits decreasing power. Something of similar is also valid for a different reactor such as an E-Cat, of course **not involving the same mechanisms** of a traditional fission reactor but different ones.

So, according to my explanation, the "Cat" would be a **subcritical reactor**, i.e. a LENR reactor that produces excess heat without achieving criticality. Instead of a sustaining chain reaction, a subcritical LENR reactor uses **additional "stimulation" from an outside source**. The source can be another subcritical LENR reactor: in our case, the "Mouse". The **LENR coupling** between them – Cat & Mouse – makes the entire system supercritical.

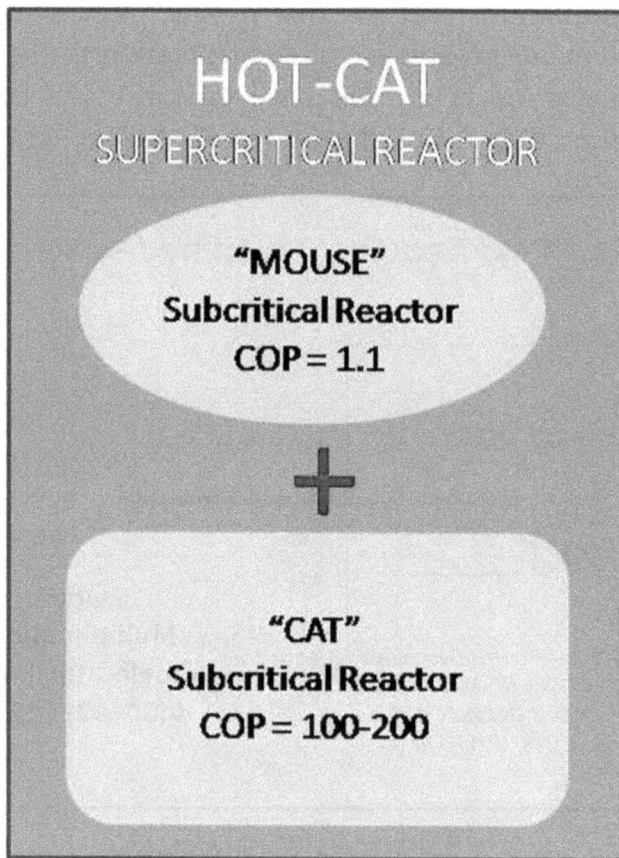

Fig. 12.2 - Model of a Hot-Cat with "Cat & Mouse" technology. When the Mouse is "ON", in this model the Hot-Cat behaves as a supercritical reactor.

The main advantage is **inherent safety**. Subcritical design of the "Cat" means that the reaction could not run away. The same is true for the "Mouse": its reactions **cannot run away** when the power input is off. Even if for some reason the heating resistor remains "on" triggering runaway reactions – or if anything went wrong – at 1450 °C **nickel would melt** and the LENR reactions would stop, then the reactor would cool down.

A comparison with conventional nuclear reactors

It is interesting to make a comparison with conventional nuclear reactors. Their nuclear fuel possesses **self-regulating properties** such as the "Doppler effect" or "void effect", which make these nuclear reactors safe. In addition to these physical properties of conventional reactors, in the nuclear subcritical reactor – whenever the external neutron source is **turned off** – the fission reaction ceases and only the decay heat remains.

In the conventional nuclear *subcritical* **reactors**, the external neutron source can be a nuclear fusion machine or a particle accelerator. Besides its application for nuclear waste incineration, there is interest in this type of reactor because it is perceived as inherently safe, unlike a conventional nuclear critical reactor, for which there exist circumstances in which the rate of fission can increase rapidly, damaging or destroying the reactor.

Instead, in the conventional nuclear *critical* **reactors**, the reactor is operated to maintain its "criticality" (the number of secondary neutrons triggering effectively a new fission) at a value of 1. Should any temporary destabilization take place, the reactor is designed to behave in such a way as to **return to the desirable criticality**.

In designing critical reactors, the nuclear engineers chose a well-defined **equilibrium point** to achieve a natural regulation of criticality, so that small variations in power naturally would correct themselves accordingly to the laws of physics. With a well chosen operating point, a variety of phenomena result in a **self regulation mechanism**.

The most important self-regulating mechanism is the "**Doppler effect**". An increase in temperature increases the thermal agitation of the nuclei present in the reactor core. It turns out that the more this thermal agitation is, the more the losses by capture of neutrons during the slowdown are. As a result, the **chain reaction slows down** and the temperature gradually drops.

In the case of a limited divergence of the chain reaction, the control of the chain reaction is facilitated by a phenomenon which may seem like a side-effect. A small fraction of the secondary neutrons are emitted with a **delay of several minutes**. If the criticality is slightly above 1, the delayed arrival of these neutrons gives time to react to prevent an explosive divergence and to introduce **control rods** absorbing neutrons.

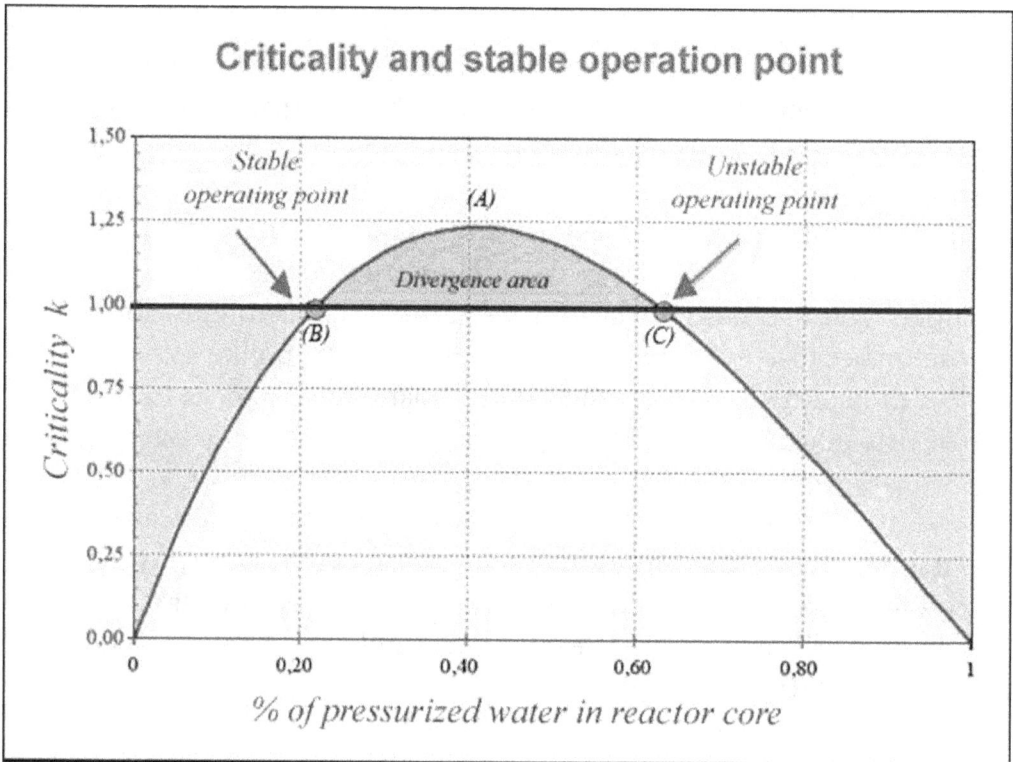

Fig. 12.3 - In a pressurized water nuclear reactor, water is used both as a moderator and a coolant. The operation point (B) is chosen by engineers because if the reactor enters into the zone of divergence of the chain reaction, the temperature increases and the amount of water (moderator) decreases, more neutrons are lost, and this brings the criticality back to 1.

The mere fact that a conventional nuclear reactor is critical or supercritical does not guarantee that it contains any free neutrons at all. On the other hand, at least one neutron is required to "strike" a chain reaction. Therefore, most nuclear reactors include a "starter" **neutron source** that ensures there are always a few free neutrons in the reactor core, so that a chain reaction will begin immediately when the core is made critical.

To improve fission probability and enable a chain reaction, uranium-fueled reactors must include a **neutron moderator** that interacts with newly produced fast neutrons from fission events to **reduce their kinetic energy** from several MeV to thermal energies of less than one eV, making them more likely to induce fission.

This is because the ^{235}U – the **typical fuel** used in conventional nuclear reactors – is much more likely to undergo fission when struck by one of these thermal neutrons than by a freshly produced neutron from fission. Neutron moderators are thus materials that slow down neutrons. Neutrons are most effectively slowed by colliding with the **nucleus of a light atom,** such as hydrogen, beryllium and carbon.

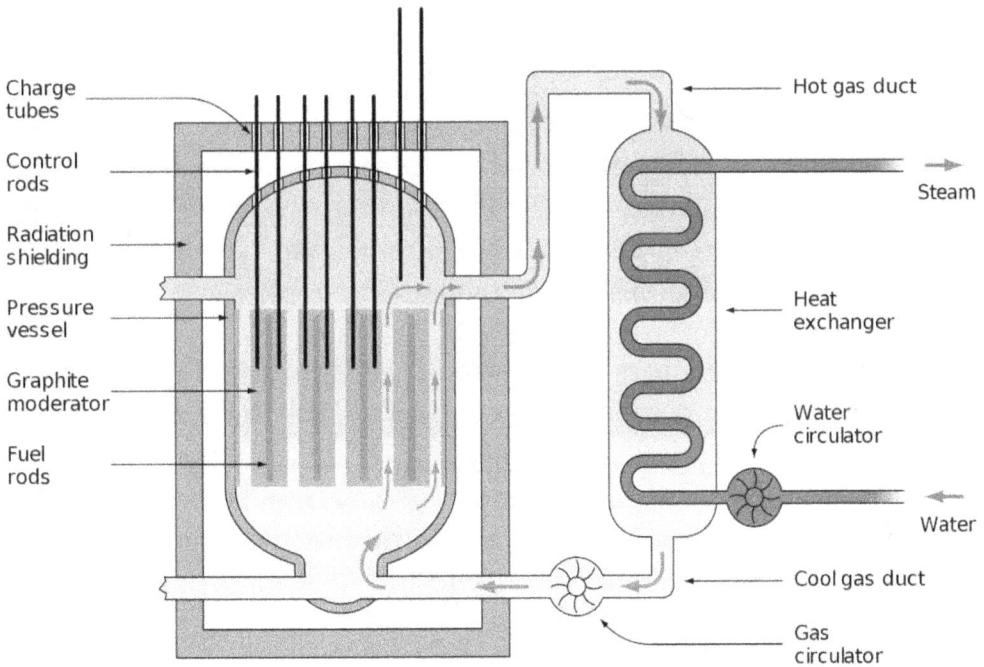

Fig. 12.4 - Scheme of a conventional nuclear reactor.

The so-called "reactivity" of a nuclear reactor is an expression of the **departure from criticality**. Positive reactivity causes power to rise exponentially at a rate proportional to the reactivity. **Negative reactivity** causes power to decrease.

The rise in temperature which accompanies a rapid rise in power generally changes the reactivity. The "**temperature coefficient of reactivity**" measures the effect of changing temperature on reactivity. Because power stability requires a rise in temperature to decrease the reactivity, a "**negative** temperature coefficient" **is necessary** for stability.

The main coupling agent in a reactor system is the **temperature field** inside the fuel, the core and its immediate surroundings. The temperature affects the neutron behavior, both in normal operation and during the transients, through the cross-sections, which account for the probability of neutrons to interact with matter in every zone and at any time.

Thus, temperature is a parameter always relevant to **normal operation and control** of a conventional nuclear reactor, but it may become extremely important and sometime decisive in the transients, mainly the reactivity driven ones, which are characterized by **very short response-time** and severe power variations.

Most moderators become less effective with increasing temperature, so *under-moderated* **reactors** are stable against changes in temperature in the reactor core: if the core **overheats**, then the quality of the moderator is reduced and the reaction tends to **slow down** (there is a "negative temperature coefficient" in the reactivity of the core).

Over-moderated **reactors** are unstable against changes in temperature (there is a "positive temperature coefficient" in the reactivity of the core), and so are **less** inherently **safe** than under-moderated cores. The amount and nature of neutron moderation affects reactor controllability and hence safety: moderators **both slow and absorb neutrons**, so there is an optimum amount of moderator to include in a given geometry of reactor core.

Any element that strongly absorbs neutrons is called a **reactor "poison"**, because it tends to shut down an ongoing chain reaction. Some reactor poisons are deliberately inserted into reactor cores to control the reaction; boron or cadmium control rods are the best example. Many reactor poisons are produced by the fission process itself, and **buildup of** neutron-absorbing **fission products** determines the **lifetime** of nuclear fuel in a reactor.

Towards a model for the "Cat & Mouse" Hot-Cat

Now you should have many ideas to imagine a model of how a "Cat & Mouse" Hot-Cat works. I want here illustrate the **most obvious** possibilities, letting to the reader the fun of exploring their details or of proposing additional ideas.

I would just remind you that, according to the already mentioned provisional patent, "as far as **Activator** is concerned, it is in turn **made up of powders** – or it makes use of them – to run the complete system". I remind you also that, as discussed at the beginning of this chapter, some kind of "LENR coupling" between "Mouse" and "Cat" (individually subcritical reactors) makes the entire system supercritical.

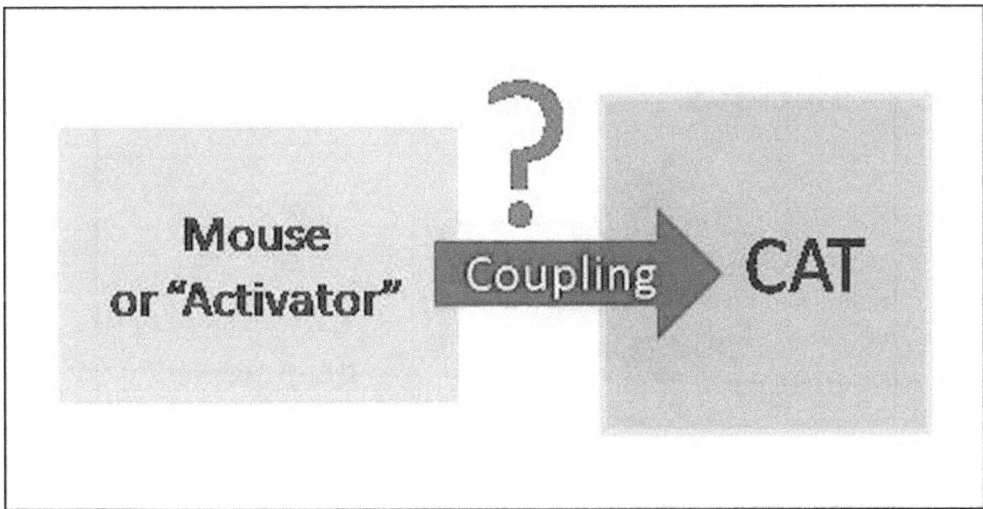

Fig. 12.5 - The interesting question about the type of coupling between "Mouse" and "Cat".

Well, I can imagine, fundamentally, **four types** of different possible **mechanisms** to couple "Mouse" and "Cat", not being the thermal one – or "trivial" solution of the problem – a real possibility for the reasons already discussed in Chapter 10 (in particular, the admission by Rossi himself in a communication with a friend of mine):

1) Low energy gamma rays

We know, from the many **specific documents** alleged to the Third Party reports describing the various Hot-Cat tests, that no neutrons, other elementary particles or nuclei, X-rays, gamma rays – or other possible products of the LENR reactions occurring inside the reaction chamber – are emitted outside the reactor, which thus has **no special shielding**.

However, in September 2012, Rossi declared in the Q&A session of a public presentation of the E-Cat technology made in Zurich, Switzerland, that the **main mechanism** for the production of excess energy in the E-Cat is through the production of **low energy gamma rays,** between **50 keV** and **100 keV** of energy. Therefore, these gamma rays are easily shielded from the outer shell of the reactor and are not detectable outside.

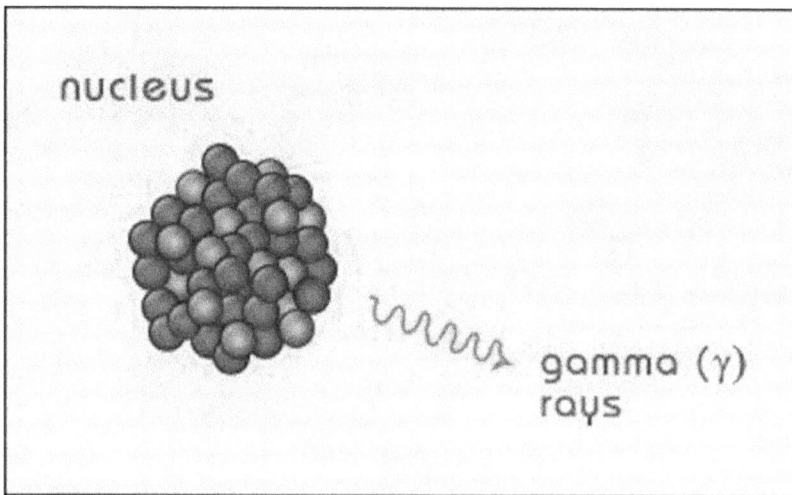

Fig. 12.6 - Gamma rays are produced by the decay of atomic nuclei as they transition from a high energy state to a lower state, but may also be produced by other processes.

It is a precious information, because these gamma rays are **photon**s that can pass from the "Mouse" to the "Cat" if the thickness between them is small enough. So, such additional photons provided by the "Mouse" photon source could perhaps **allow to control a "Cat"** as other particles do in a conventional nuclear reactor, the difference being that source-driven photons are easily controlled by the electrical power input.

2) *Electromagnetic fields*

An alternative mechanism of coupling between "Cat" and "Mouse" is provided by the **electromagnetic fields**. We know that a variable (pulsating) electrostatic field is produced inside the E-Cat and that it can be detected outside the reaction chamber, as illustrated in Chapter 9, dedicated to the salient features of a Hot-Cat.

Static electric fields store energy and can influence conductors and moving charges, but they emit no radiation. Variable electric and magnetic fields are **interdependent**. For example, an alternating field of one type produces an alternating field of the other type, with the same frequency. In this way, some of the energy produced radiates away from the source, and higher frequencies have **greater ratios** of radiated to stored energy.

More exactly, **low frequency** electric and magnetic fields can exist separately, because at these frequencies the shifted currents are negligible. But a rapidly changing electric field cannot exist alone: it is associated to a **variable magnetic field**, thus both generate a variable electromagnetic field. Being the electrostatic field described by Rossi as "**pulsating**", weak (pulsating) electromagnetic fields should be induced.

3) *Proprietary waveforms*

From Chapter 2 of this book, you should remember that in the Hot-Cat "Type II" – powered with a constant power input, not with a "Cat & Mouse" self-sustained mode – "the resistor coils were fed by a **Triac power regulator** which interrupted each phase periodically, in order to modulate the power input with a controlled waveform, which is an industrial trade **secret waveform**. This procedure was needed to **properly activate** the powder reaction charge".

In the Chapter 3, we saw that in the Hot-Cat "Type III" – powered with a "Cat & Mouse" self-sustained mode instead of a constant power input– was used a **single-phase power supply**: that is, a **controller circuit** having a three-phase power input and single-phase output, within a housing whose contents were not available for inspection, inasmuch as they are part of the **industrial trade secret**.

Fig. 12.7 - The Compact FUSION three phase power controller, a digital controller used in the test performed on the Hot-Cat.

Also in the Lugano test described in Chapter 4 – where the Hot-Cat was powered with 2 different constant power inputs for about 10 and 20 days, respectively – the resistor coils were fed with some **specific electromagnetic pulses** covered by industrial trade secret. So, electromagnetic pulses or, more in general, **controlled waveforms at proper frequencies** could be applied at the same time as the power input and used to control the reactor.

We do not know their function: perhaps, in this way you have one **mechanism stimulating** the nuclear reactions in the "Mouse" and/or in the "Cat" and another mechanism controlling them, simultaneously. However, such as mechanism cannot represent the "Mouse" – because from the provisional patent we know that "the Activator (or Mouse) is made up of **powders** or makes use of them" – so must be excluded.

This kind of stimulation is probably **an evolution** of that already used in 2011 during the heating or "activation" phase of the "low temperature" E-Cats, i.e. until the self-sustained mode began. It is well described by Mats Lewan in the public test of an E-Cat he carried out on October 6, 2011: Lewan writes that at 15:53, 4 hours after the heating resistor had been turn on, **"a device 'producing frequencies' was switched on"**.

Fig. 12.8 - The control box always used by Rossi to power the "low temperature" E-Cats. The panel shows two heating resistors switched on.

Moreover, low temperature E-Cats were fed with a 230 V single-phase power input **in an intermittent way**, through a blue control "box" containing 12 phase fired controllers of the type "**VL676**", produced by company **GSEI** in Genoa, Italy. Such controllers can be, depending on type, regulated by **10 or 20 A** in 20 half-steps (0..9) and have a consumption of 2 watts (in sum 24 W) from a **24 V** voltage source.

When during a test performed by Mats Lewan on September 7, 2011, the just mentioned control box was (at 19:00) switched on setting to the value "5", according to Lewan's words "the system apparently switched on and off the power **intermittently** about every second, which resulted in an input AC current that went continuously between zero and 11.4 A.

During the same test, when the power (at 19:10) was raised to "6", the power was still switched on and off, **the interval with power on was slightly increased** and the interval with power off was decreased. When the power (at 19:20) was raised to "7", and later (at 19:30) to "8", the interval with power on was increased further. When the power was increased to the maximum value, "9", power was at this point **constantly switched on**.

4) Substances released

The previous types of coupling are all examples of ***remote* coupling**, i.e. they assume that "Mouse" and "Cat" are physically separated, preventing mixing of their different powders and of the produced gases. However, if you remove at least partially this restriction, then it is possible a new mechanism: a direct, **physical coupling**, through the release of one or more substances in the reaction chamber as the temperature increase.

For example, a **catalyst** – or the hydrogen itself – could be added by direct heating through the resistor coils. The idea behind this mechanism would be that of increasing, at the right moment, the **fuel concentration** or the catalyst concentration to pass from a subcritical regime in a supercritical state. This assumes that we are somehow able to control the concentration of one of the reactants or of the catalyst.

Unfortunately, this seems not to be our case, because it is difficult to imagine how a released substance **can be absorbed again**, if not after a significant lowering of the temperature in the entire reactor.

Moreover, we know that "Mouse" and "Cat" use different powders and devices (Rossi on JoNP: "we separated the **activation** from the **reaction in 2 separate devices**"), so that a physical contact among the powders or the gases – and thus, consequently the proposed mechanism – must be surely excluded.

CHAPTER 13

TWO UNDERESTIMATED PROBLEMS

The **tests** of a Hot-Cat "Type IV design", or "Dog Bone", described in TPR-2 are very interesting because, instead of providing answers further clarifying the operation mode of such a device, **created new questions** and problems not easy to solve.

The attention of many focused on the **surprising results** of the analysis regarding fuel and ash used in such a test, performed at specialized Swedish laboratories. But, in my opinion, this is not the greatest mystery linked to the test, because I believe that at least one of the two powder samples analyzed is the result of a **deliberate misdirection**, which would explain some oddities emerged from the analysis itself.

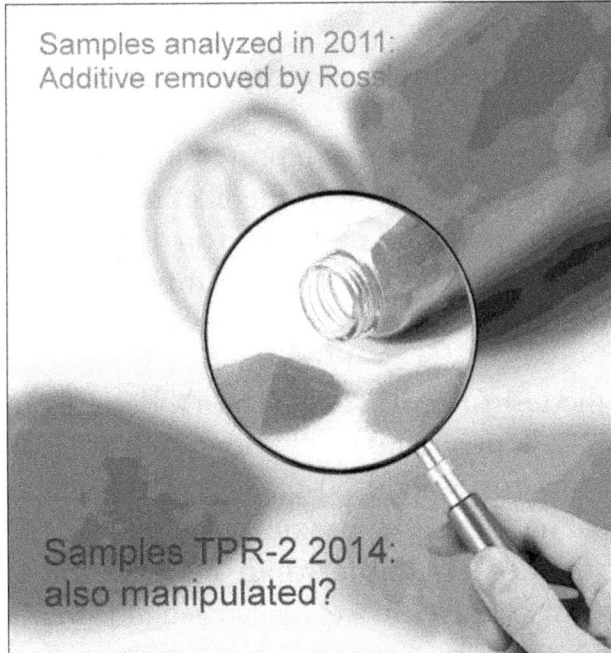

Fig. 13.1 - The misdirection performed in 2011 casts shadows on the analysis of the powders of fuel and ash made in 2014 and attached to TPR-2.

This should not surprise. Already in 2011, **samples of fuel and ash** powder were provided to Prof. Sven Kullander by Rossi to be analyzed, and the results were **in total contrast** to the more recent ones, and also with the laboratory analysis mentioned, in 2010, in the article by Rossi and Focardi "*A New Energy Source from Nuclear Fusion*".

Recently, Rossi **admitted** that these powders analyzed in 2011 had **certain additives removed**. So, if at the time – probably to slow down the work of the possible competitors – a deliberate misdirection was made by **altering the composition** of the powders provided for analysis, there is no logical reason for which the same thing **did not occur again** with the powders of the "Dog Bone", making their laboratory analysis useless.

Instead, there are two other "mysteries" – or at least **major problems** – concerning the tests of Rossi's Dog Bone that have been largely **underestimated** and that, on the contrary, deserve great attention. Given their importance, I will now illustrate them in some detail.

The problem of the high internal temperature

If we assume that the temperatures reported in TPR-2 are correct – and MFMP's experiments on a Dog Bone-like reactor showed that the delicate thermal measurements performed in Lugano with IR cameras are correct – there is a big problem: **the temperature at the center of the reactor is too high** for the survival of nickel.

Indeed, during the test of an E-Cat "Type IV design", or "Dog Bone", described in TPR-2, a power input of about 900 W has been applied to the heating resistors for around 20 days, and the temperature **outside** the reactor in this period was approximately 1400 °C. The temperature of the core was then **surely higher than 1453 °C**, the melting point of the nickel, and the nanometer nickel melts even at lower temperatures.

In TPR- 2 there is **no information** on the central temperature of the reactor, perhaps as there is not in that part of the device a thermocouple measuring it. But we can get an idea of what was the temperature at the centre of Rossi's Dog Bone watching what was the central temperature of **MFMP's Dog Bone at the same power input** (900 W).

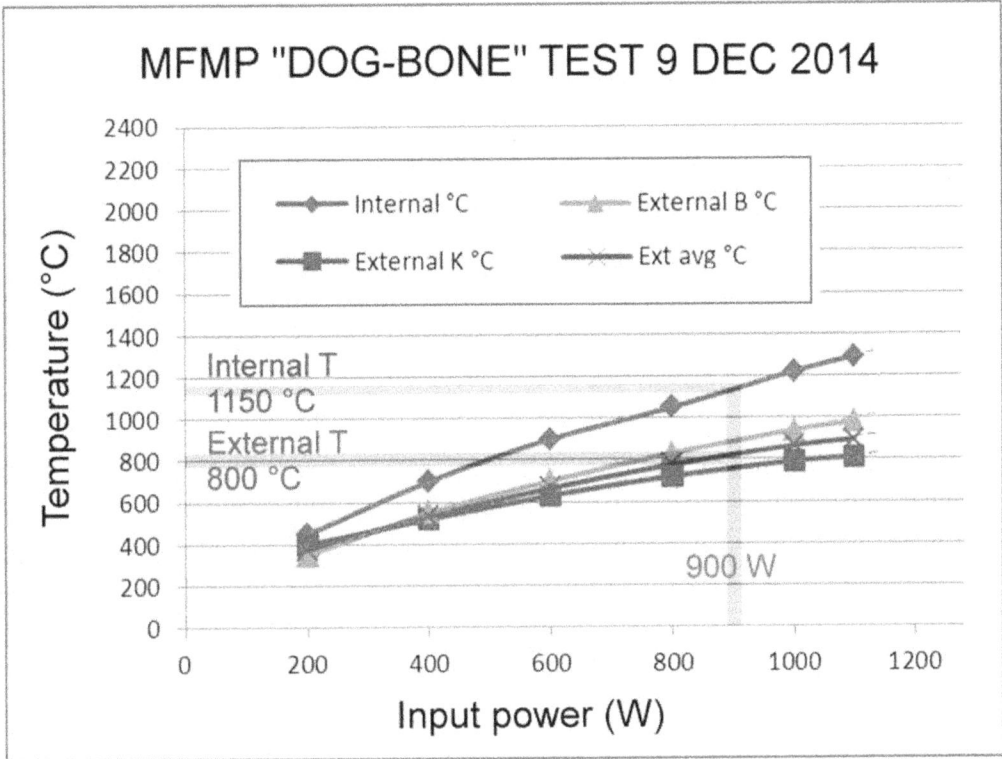

MFMP "DOG-BONE" TEST 9 DEC 2014

Internal T
1150 °C

External T
800 °C

900 W

Fig. 13.2 - Temperature vs. power input curves for a "Dog Bone"-like reactor tested by MFMP on December 9, 2014, but not well sealed and thus working such as a "dummy".

From the chart of the MFMP experiment performed on December 9, 2014, we can see that the central temperature should be **about 1150 °C**, and the external temperature about 800 °C. So, in MFMP's Dog Bone, the **difference between external and internal** temperature is around 350 °C. Assuming that in Rossi's Dog Bone the difference is similar, the central temperature in such Hot-Cat would be (see Fig. 13.3) **1750 °C**!

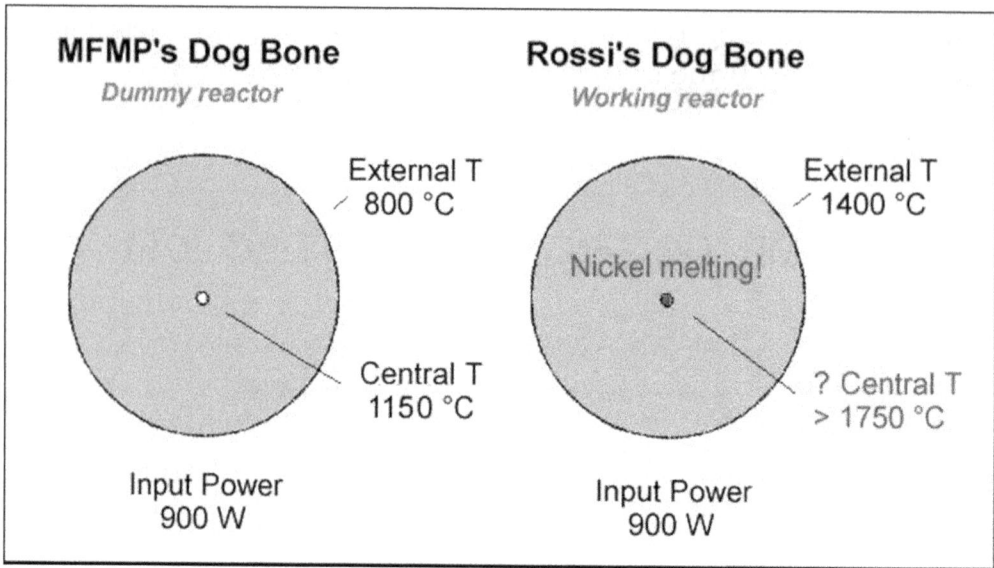

MFMP's Dog Bone
Dummy reactor

External T
800 °C

Central T
1150 °C

Input Power
900 W

Rossi's Dog Bone
Working reactor

External T
1400 °C

Nickel melting!

? Central T
> 1750 °C

Input Power
900 W

Fig. 13.3 - The central temperature in the MFMP's and Rossi's "Dog Bone" reactors.

You can also calculate the thermal delta by yourself using a **dedicated software**: for example, the free interactive tool "Steady-State Temperature Profile of Two-Layer Pipe" provided by *Wolfram Project*. This tool shows the temperature profile of a **two-layer pipe with constant interior and exterior** surface temperatures. You can vary the radii of the two material layers, their respective conductivities, and the two surface temperatures.

In the example here published, you can see the resulting temperature profile as a function of pipe radius, with a **slope discontinuity** at the **interface** of the two materials: with the parameters chosen in our example, the difference between the internal and external temperature is 350 °C, the same value found by MFMP. By minimizing the conductivity of one of the materials, an **insulating layer** can be simulated.

This approach uses the concept of **thermal resistance circuits**. Thermal resistance for a cylindrical shape as a function of radius r is given by $R(r) = Log(r / R_{max})/k$, where R_{max} is the maximum cylinder radius and k is the thermal conductivity. Heat transfer per unit length using thermal resistance is given by $q' = 2\pi\Delta T / R$, which is then used to solve for the temperature as a function of radial distance r.

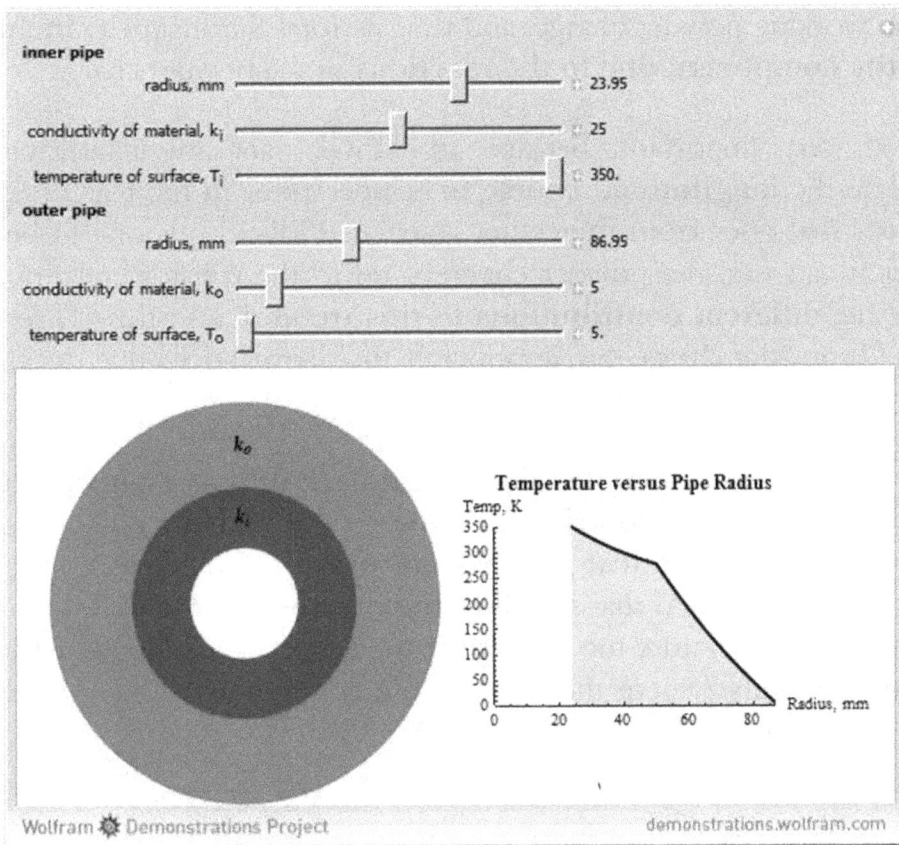

Fig. 13.4 - A decreasing temperature profile as calculated with the tool of Wolfram Project.

Our previous estimate is indirectly confirmed by Penon's test on a Hot-Cat discussed in Chapter 6. In such a case, an average Hot-Cat module temperature peak of 801° C was measured through an IR camera, whereas the **temperature of the internal cylinder** – which may be seen to be glowing white hot in the photographs – ranged **from 1100 °C to 1200 °C**, and was seen to increase from the entrance of the hole towards the center of the cylinder. This last measurement was performed with a **laser-guided** thermometer.

So, also in the Hot-Cat tested by Eng. Penon the difference between external and internal temperature, or **thermal delta, was 300-400 °C**, that is around 350 °C, in perfect agreement with our estimate made through MFMP's experiment on a dummy-like Dog Bone, the difference being that in Penon's test the reactor was not a dummy but contained an active, working powder charge, and thus its total thermal delta **included also the component due to the reactions** in the powder charge.

This is very important, because in TPR-2 data are insufficient to calculate the **longitudinal course** of temperatures in the reactor, i.e. in the area that goes from the center – where TPR-2 leads one to believe that you can find the powder charge – up to the resistor coils, nor we know the **different contributions to this trend** due to the: (1) resistor coils; (2) powder charge. So, in principle the thermal delta due to (2) may have **offset** that due to (1).

But, as we've just seen, the chart of the MFMP's test provides clues about (1), that is the **contribution due to the resistor coils**, whereas Penon's test indicates that (2) does not offset (1). Therefore, it is not difficult to understand that **much of the reactor volume** in Rossi's Dog Bone has a temperature too high, so that only **very close** to the electrical heating elements (where the temperature is about 1400 °C) nickel can hope not to melt.

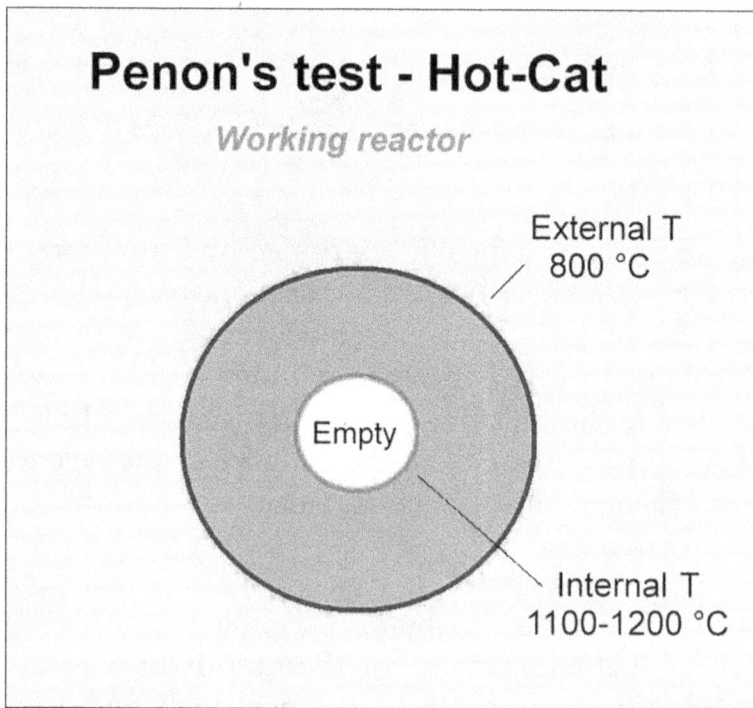

Fig. 13.5 - The temperatures of the external and internal cylinders of a Hot-Cat in Penon's test.

In conclusion, the **only three reasonable explanations** of the "mystery of the high internal temperature" emerged from Rossi's Dog Bone reactor **are the following**: (1) there was not charge in the reactor; (2) the charge was not in the central part of the reactor and also in much of the reactor volume between the central 4 mm wide tube and the resistor coils; (3) the external temperature of the reactor **was not the one** indicated in TPR-2.

As we discarded the third possibility at the beginning of this section thanks to MFMP's work, and also the first having been shown in Chapter 7 that a Hot-Cat really works, only one explanation remains: the (2)! In other words, **the charge could be only very close to the resistor coils**, if not on the surface of the resistor coils themselves.

The problem of the powder distribution

Another major problem related to the tests illustrated in TPR-2 concerns the **distribution of the powder charge inside the reactor** as, from some photos in the report showing the Hot-Cat glowing, the powder seems to be distributed in a **perfectly uniform way**, whereas considering how it has been inserted it should not be at all!

Indeed, we read in TPR-2 (pages 3 and 7): "The hole for the thermo-couple probe is also **the only access point** for the fuel charge. [..] The powder had been previously placed in a small envelope, weighed (about 1 g), and then transferred to a test tube so that Bianchini could perform radioactivity measurements on it, after placing it in a low background lead well. Lastly, the contents of the test tube **were poured inside the reactor**, in the presence of a member of the experimental team. The leads were reconnected and the cap sealed with a mixture of water and alumina powder cement".

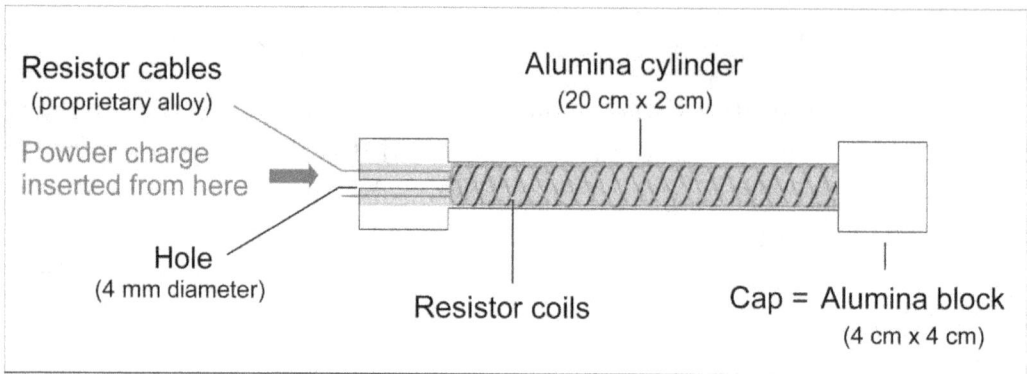

Fig. 13.6 - The hole for insertion of the powder charge in a "Dog Bone", according to TPR-2.

The device was about 12 inches long, and the "fuel" was poured into a 4 mm hole. How was the "fuel" **evenly distributed** along this 12 inch, 4 mm (or larger) cavity? This is not a trivial question! They had to hold the device upright to pour the powder into that very small 4 mm hole. Then they placed it horizontally, as we see in all report's pictures.

Remember that **the device is opaque,** so they could not see the "fuel" inside the alumina cylinder. And also consider that 1 gram of powder is a very small amount. But yet the photos published in the report let us to believe that bright glow which is **along the full length** of the device is caused by the "fuel" glowing from some kind of LENR process.

Indeed, the authors themselves say in TPR-2 (page 25): "The resistors appear **to glow intensely** in the parts lying outside the caps, whereas inside the reactor body they seem to shade an underlying emission of light. This may be explained if we consider that **the main source of energy** inside the reactor body is actually **the charge,** and that it is emitting more light than the resistors".

It's interesting that the authors also write in their report, about the calculations of the fuel's energy density: "Considering that **we do not know the internal structure** of the reactor, and therefore cannot completely rule out that there were **other charges inside it** besides the one weighed and **inserted by us,** we may repeat the above calculations taking the weight of the entire reactor (452 ± 1 g) into consideration…".

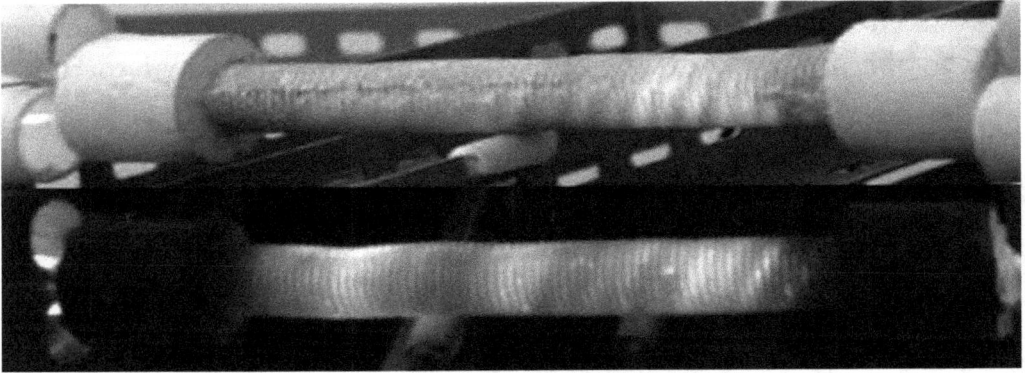

Fig. 13.7 - Rossi's "Dog Bone" glowing during the Lugano test (from TPR-2).

Here the authors are saying in the report they are the ones who "inserted" the "charge" or "fuel" into the device for the real test. But the authors also say in the same report: "**Rossi later intervened** to switch off the dummy, and in the following subsequent operations on the E-Cat: **charge insertion**, reactor startup, reactor shutdown and powder charge extraction". In any case, our initial problem remains.

Indeed, let suppose that after putting the reactor vertically and having poured the powder inside, the reactor has been picked up and **shaken to distribute** the powder uniformly. Even assuming that the inner wall of the chamber containing the powder was highly porous so that the powder attacked, we **never** would have been able to obtain a uniform distribution along the tube of the reactor and along its circumference.

It's not my field of expertise, but it is possible to imagine that, due to the **temperature gradient in the air** of the chamber containing the powder and the hydrogen gas released from the powder itself, there would be established **convective motions** that circulated nickel longitudinally, bringing it from time to time in the central area, where surely **it would have to melt** for the reasons outlined in the previous section of this chapter.

In conclusion, the **only three reasonable explanations** of the "mystery of the uniform distribution of the powder" are, in my opinion, the following: (1) there was not charge in the reactor; (2) on the walls of the 4 mm tube (or of a wider tube) there was already the "real" powder charge – or its solid components – evenly distributed; (3) the "real" powder charge **was already** evenly distributed in a reaction chamber near to the resistor coils, if not **on the surface of the resistor coils** themselves.

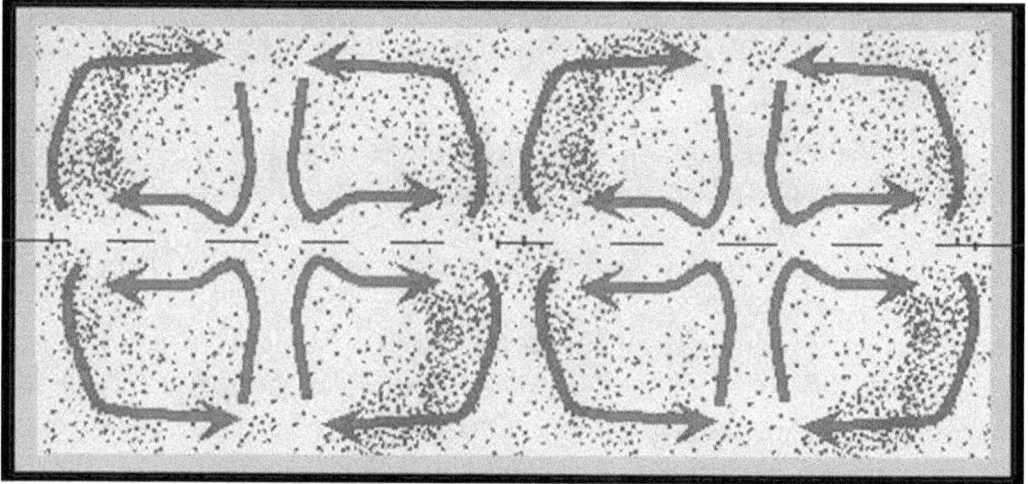

Fig. 13.8 - Possible convective motions that distribute the powder in the reaction chamber.

We can **exclude the first possibility** having been shown in Chapter 7 that a Hot-Cat really works. The remaining two possibilities **seem both acceptable**, but if we take into account the conclusions of the previous section we know that the real powder charge could not survive in the central 4 mm tube, so or (a) the tube beyond the caps was much wider than 4 mm or (b) **the only possible solution** of the problem is the number (3).

CHAPTER 14

A SOLUTION FOR THE POWDER CHARGE

In the previous chapter we realized that the reactor described in the Lugano report is actually a **rather surprising device**, because you must somehow explain the mystery of the high internal temperature that makes impossible the survival of nickel in much of the reactor volume and that of a uniform distribution of the powder charge inside.

But perhaps even more surprising is that, if we analyze carefully the previous official tests of the Hot-Cat, we find even in such cases **apparent oddities**.

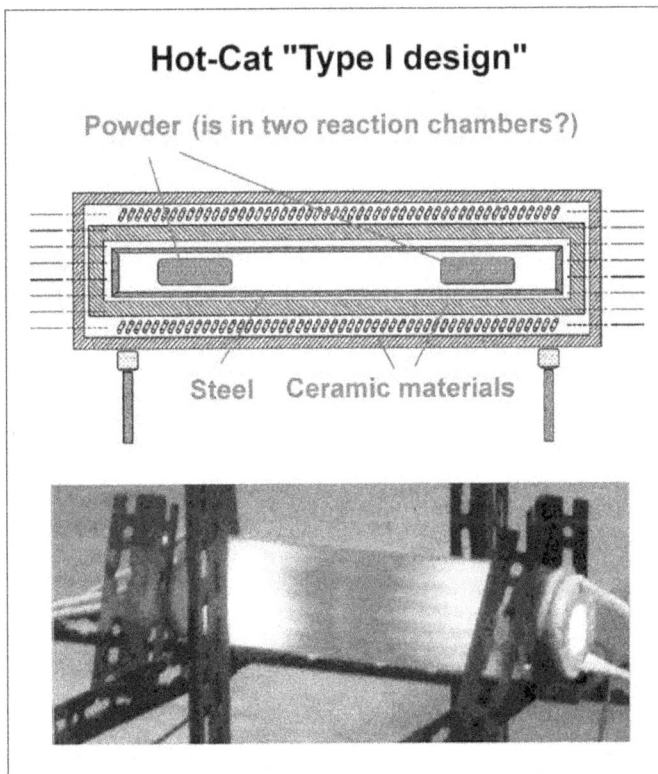

Fig. 14.1 - Charges and materials in the Hot-Cat "Type I design".

For example, in the Hot-Cat "Type I design", the two powder charges are inside a steel cylinder, but being **placed laterally** along the same. Thus, it is likely that they are actually placed in **two containers** (of unknown material) housed inside the steel cylinder, since they could hardly be "**attacked**" somehow uniformly to its surface.

In this reactor, the steel cylinder which contains the charge is enclosed in a larger cylinder of a **not specified** ceramic material, perhaps different from those in which the heating resistors are likely to be immersed. Looking at the glowing reactor, however, we see that the light of the side charges is **not concentrated** near the central axis of the reactor but is distributed along its diameter, which is also compatible with a not central charge.

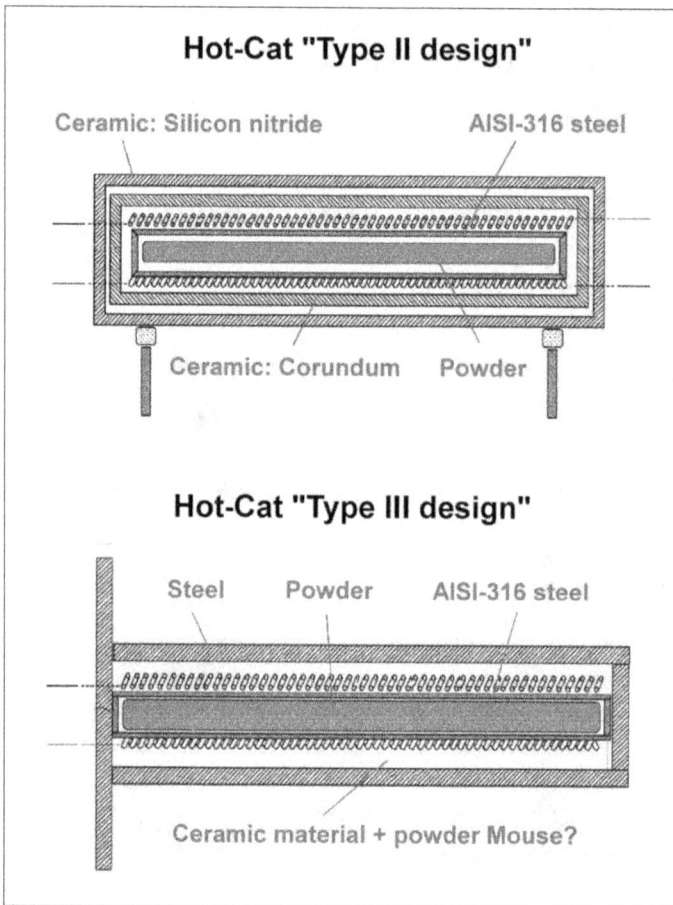

Fig. 14.2 - Charges and materials in the Hot-Cat "Type II and III design".

Instead, in the Hot-Cat "Type II design" things seem clearer. The charge is uniformly distributed, in some way, in the central cylinder of steel, and the heating resistors, **possibly immersed in a ceramic material**, are surrounded by a cylinder made of corundum, material different from that of the outer cylinder, which is silicon nitride.

In the Hot-Cat "Type III design" – the self-sustained reactor with the "**Cat & Mouse**" technology – the powder of a "Cat" is located in the steel cylinder and the heating resistors are likely immersed in a ceramic material, but it's not clear where you can find the **2nd charge**, that of the "Mouse", which we know to be physically separated from the other.

Finally, in the reactor of the Penon's test – that is prior to the reactors described so far – the **powder is not in a steel cylinder**. According to the engineer "Cures", the charge would be in a reaction chamber located between the internal and external cylinders. This means that there is an **"ad hoc" reaction chamber** or that the powder is in the **empty spaces** in which the resistor coils are housed. In the latter case, however, it is difficult to imagine how it can be evenly distributed if it is not attached to the same resistors.

Type of problem	Penon	Type I	Type II	Type III	Type IV
High central temperature	Ok	Ok	Ok	Ok	Yes
Not uniform distribution	Yes	Ok	Yes	Yes	Yes
Volume for gas expansion	Yes	Ok	Ok	Yes	Yes
Chamber not recognizable	Yes	Yes	Ok	Yes	Yes

Tab. 14.1 - A table resuming the main types of "problems" in the 5 Hot-Cats officially tested. "Yes" indicates that the problem is unsolved.

Therefore, at this point it should be clear that the problems of where the powder charges are and how the powder is uniformly distributed are two major unsolved questions that we encounter **in more than one reactor**, as summarized in the table. The question now is the following: can we find **a common solution** to these problems?

Solution 1: The "doped" resistor coils

One interesting possibility is that the powder – or at least its components that remain solid even at high temperatures, such as nickel and perhaps the secret catalyst – is **included in the heating resistor** itself, or at least covers its surface.

After all, as Andrea Rossi himself admitted on his JoNP, some days after the release of the 2nd Third Party Report describing the Lugano test, "the coils of the reactor are made with a proprietary alloy, and the Inconel is only a **component doped** of it" (Inconel is composed of a high percentage of nickel in addition to other elements). Perhaps, it's doped to increase resistance to corrosion or operating temperatures.

However, this suggests that not only the Mouse portion of a Hot-Cat could be the doped, high temperature grade Inconel resistor wires, but also that in reactors without a mouse the wires could perhaps be **part of the reaction chamber**, hosting at least part of the powder – that is some of the components of the charge – **on their surface**.

This would solve, in principle, the already described "problem of the central temperature" in the Lugano test, because the reactions would take place not in the central part of the reactor but on the surface of the heating resistors, and **nickel would not move freely** throughout the reactor, risking ending up in the most central areas, where we know that the extreme heat could melt it.

Type of problem	Penon	Type I	Type II	Type III	Type IV
High central temperature	Ok	Ok	Ok	Ok	Ok
Not uniform distribution	Ok	Ok	Ok	Ok	Ok
Volume for gas expansion	Ok	Ok	Ok	Yes	Yes
Chamber not recognizable	Yes	Yes	Ok	Yes	Yes

Tab. 14.2 - The problems solved by "Solution 1" in the 5 types of tested Hot-Cat: they are indicated with "Ok", whereas the still unsolved with "Yes".

Furthermore, it would also solve the problem of the uniformity in the distribution of the powder charge, since the **solid matrix of the nickel**

and/or of the catalyst may be **evenly distributed on the surface** of the heating resistors before resistor coils being mounted on the reactor, possibly at least in part embedded in a ceramic material.

Finally, it would also explain why, in the dummy run of the reactor described in the Lugano report, the **heating was stopped** – strangely enough – **below 500 °C**, without a valid reason. One explanation could be that in the Hot-Cat tested in Lugano the real powder charge was not that added in the 4 mm wide central cylinder, but that it was **already present** in the heating resistor or on its surface.

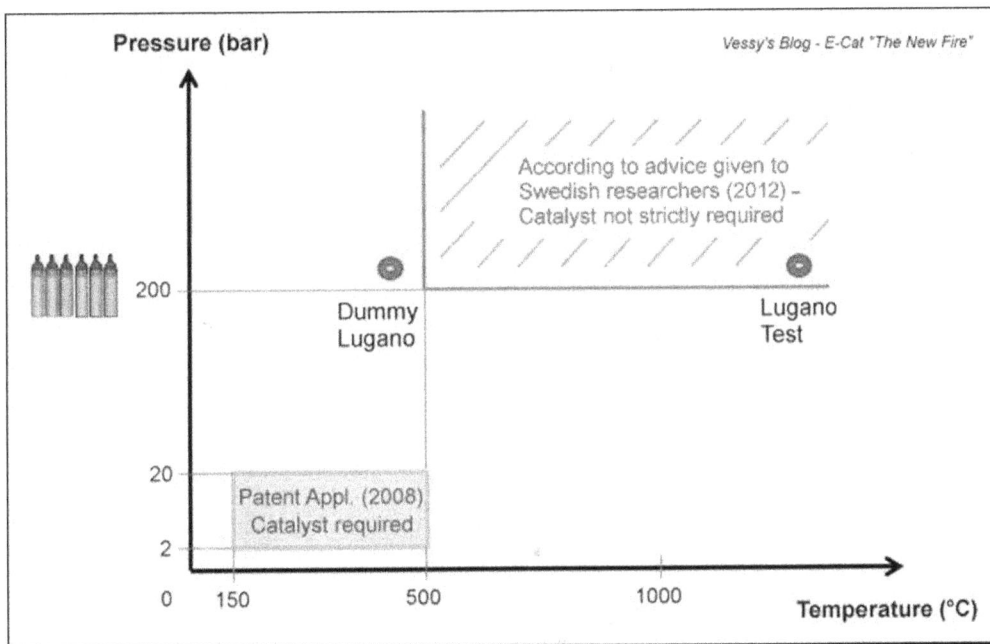

Fig. 14.3 - Why the Hot-Cat dummy was heated below 500 °C in Lugano: in this way, Rossi Effect could not take place and give excess heat.

The weak point of our solution of a charge (at least partially) contained in the heating resistor consists in the fact that this explanation requires the presence of a **small void volume around the resistor coil**, in which the hydrogen can be free to reach the necessary pressure to load the nickel lattice. And it's **still** all to show that, at least in some Hot-Cat, such space around the heating resistor **actually exists**.

So, we do not know if the previous explanation is valid for all Hot Cat-tested, however it could be very likely the **right solution** for the third

Hot-Cat tested and described in TPR-1, that of the "Cat & Mouse" technology. In this reactor, the powder of the "mouse" or Activator could perhaps be actually **deposited on the resistors' surface**.

Indeed, if you remember Rossi's reply to Georgehants (March 24, 2013): "The new generation of Hot-Cats is made by a two stage system, consisting of an Activator with resistors coupled with **a kind of charge**, which activates the Cat with a different charge". And also the provisional patent says that "the Activator is made up of powders or makes use of them", without clarification of the meaning of "**makes use**".

On the other hand, once Rossi said that "the destructive test shows what happens if we leave the self-sustained mode go ahead without driving heat with the electric resistor we call the Mouse". Then, the heat provided by the electric resistance could be useful **to moderate the reaction** of the Cat, whereas the LENR reactions produced by the powder inside the Mouse (which is a mini-reactor with COP 1.1), on the contrary, would **excite it**.

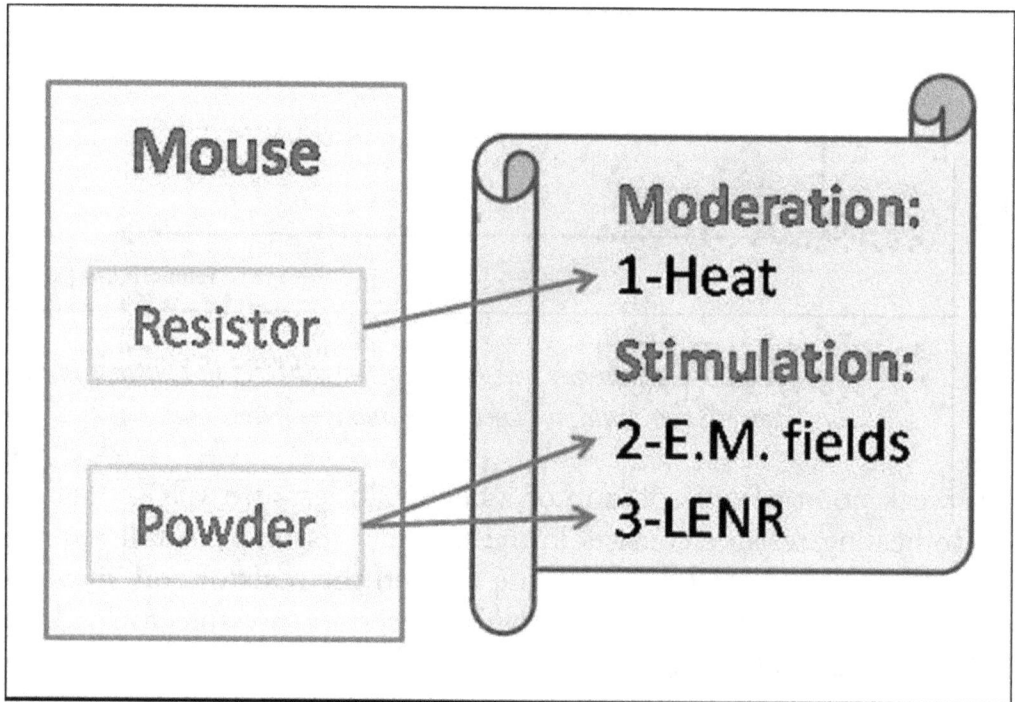

Fig. 14.4 - Model of the coupling between "Cat" and "Mouse" with this new Rossi's information.

Solution 2: The hidden reaction chamber

Another possibility is that the powder charge is contained in a **reaction chamber** created in some way **within the ceramic material**, probably as has happened in the case of the reactor tested by Eng. Penon, where according to Cures a reaction chamber would exist – together with the heating resistors – between the two internal and external cylinders.

A reaction chamber placed **near the heating resistors** – or even **around** the same resistor coils – would solve the problem of the high central temperature in the Dog Bone reactor tested in Lugano, and the problem of **doubtful results** regarding the isotopic analysis of fuel and ash, since the real powders would be hidden inside.

This solution would also resolve the case of the reactor with "Cat & Mouse" (because the Mouse would be in a reaction chamber close to the heating elements) and the question of the **volume required for the release of gas** and for achieving its ideal pressure, whereas it is not clear how a uniform distribution of the powder in the reactor can be guaranteed.

Type of problem	Penon	Type I	Type II	Type III	Type IV
High central temperature	Ok	Ok	Ok	Ok	Ok
Not uniform distribution	Yes	Ok	Yes	Yes	Yes
Volume for gas expansion	Ok	Ok	Ok	Ok	Ok
Chamber not recognizable	Ok	Ok	Ok	Ok	Ok

Tab. 14.3 - The problems solved by "Solution 2" in the 5 types of tested Hot-Cat: they are indicated with "Ok", whereas the still unsolved with "Yes".

Moreover, recently Rossi himself revealed, in an interview by Skype with Allan Sterling, that the **"Cat" has a volume of a glass of whisky** (about 10 centiliters) and has 2-3 grams of charge (nickel), whereas the Activator, or "Mouse", is **physically bigger** than the Cat because it contains the resistors, but has less charge: about **0.5 grams**.

On March 25, 2013, Rossi replied to a reader who asked whether the Mouse is **fixed inside** the Cat: "No, it is not". So, now we have an idea of how physically the system "Cat & Mouse" is done. It should however be emphasized that the expression "glass of whisky" refers to the **volume** of the reaction chamber, not to its geometry, having the charge always necessarily be distributed along the entire length of the reactor.

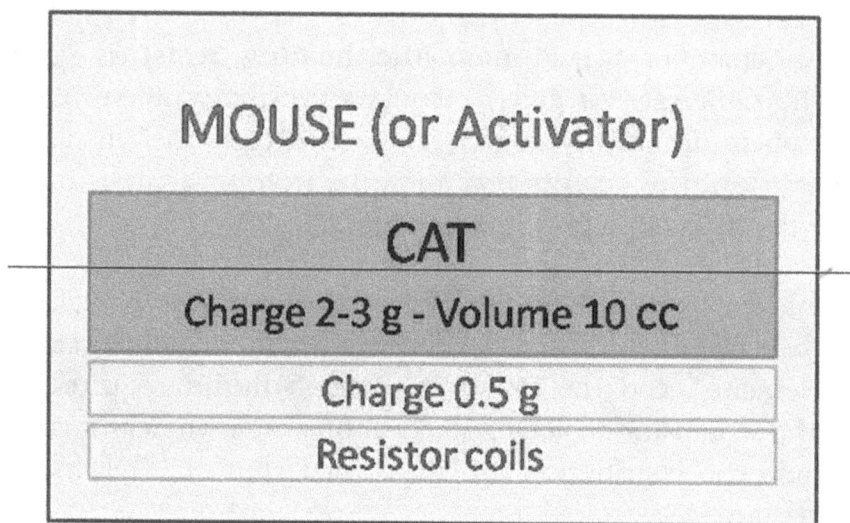

Fig. 14.5 - Model of a "Cat & Mouse" system according to Rossi's information.

In particular, since the reaction chamber – especially in the Hot-Cat tested in Lugano – should be near the heating resistors, it may be a **long empty cylinder** between the resistors and the inner part of the reactor, although the tiny reactor used in the Lugano test does not appear to have any **other internal parts** other than the helically wound resistor coils. And, for the same reasons discussed above, it would be **extremely difficult** to distribute the powder uniformly in such a chamber.

The solution of a **hidden reaction chamber**, however, would be in accordance with the recently statements made by Andrea Rossi: that the reactor tested in the Lugano report contained **both the "Mouse" and the "Cat" portions** and that in such device there was only one type of powder (in another occasion, he said that "when the start/stop or self-sustain mode is not used, **the difference** between the Mouse and the Cat **vanishes**").

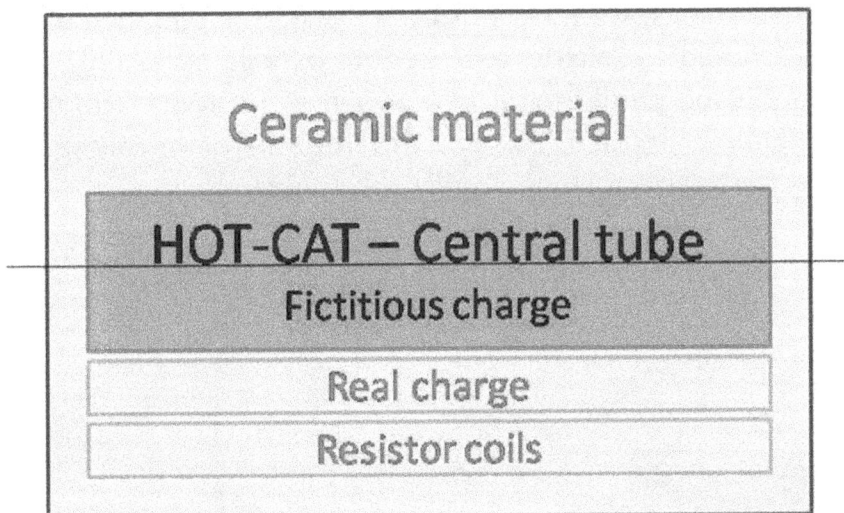

Fig. 14.6 - A model of the Hot-Cat used in the Lugano test which solves all the "problems".

Note that the problem of a uniform distribution of the powder has not been addressed by Parkhomov, so in the other tests he carried out many **destructive hot spots** originated, whereas in the best test shown in Chapter 5 things went relatively well. Therefore it is possible that, **within certain limits** and conditions still not well known, the powder is distributed fairly uniformly in the reaction chamber **simply by heating** the reactor.

Therefore, there is **no a clear best solution** between the two possibilities we have proposed, but in any case both provide that – at least in some reactors – the charge is very close to the heating resistor, if not on its surface. However, if you use such solutions – choosing appropriately one or both of them from reactor to reactor – it is possible to explain the **100% of the problems** for the 100% of the reactors.

There is little doubt, however, that a **good understanding** of the problems related to the localization of the powder charges in the reactor and of the spatial distribution of the powder within the reaction chamber **is essential** to develop a stable and reliable **second generation E-Cat** and, ultimately, a commercial product.

Among other things, we must not forget that, in a commercial product, the charge should be **easily replaceable**, which necessarily put some constraints on the number and type of solutions adoptable, unless one does not think of replacing, after some months of operating time, the **entire reactor** to control it and replace the charged.

ABOUT THE AUTHORS

Roberto Ventola is a 38 years old Italian Electrical Engineer. With over a decade of experience, he is currently working in a multinational company manufacturing electronic equipment, particularly in the computer field. He has a strong foundation in mathematics and possesses the creativity needed to develop projects regarding new technologies. His dream? To be the first to replicate a second-generation Hot-Cat.

E-mail: roberto.ventola@yahoo.com

Vessela Nikolova, psychologist, lives in Tuscany, Italy, and is already author of a book, *E-Cat - The New Fire*, the biography of Andrea Rossi. She founded and now manages a blog about Low Energy Nuclear Reactions (LENR) and the E-Cat, touching on new themes in technology, targeted at an international audience.

E-mail: vesselannikolova@gmail.com

www.ingramcontent.com/pod-product-compliance
Lightning Source LLC
Chambersburg PA
CBHW080558220326
41599CB00032B/6524